Techniques for the Organic Chemistry Laboratory

Biological Perspectives and Sustainability

Techniques for the Organic Chemistry Laboratory

Biological Perspectives and Sustainability

GREGORY K. FRIESTAD

THE UNIVERSITY OF IOWA

W. W. NORTON & COMPANY

Celebrating a Century of Independent Publishing

Editor: Rob Bellinger
Senior Associate Managing Editor, College: Carla L. Talmadge
Developmental Editor: John Murdzek
Editorial Assistant: Aidan Windorf
Associate Director of Production, College: Benjamin Reynolds
Managing Editor, College: Marian Johnson
Media Editor: Marilyn Rayner
Chemistry Content Development Specialist: Dr. Richard L. Jew
Associate Media Editor: Liz Vogt
Media Project Editor: Jesse Newkirk
Media Assistant Editor: Manny Ruiz
Managing Editor, College Digital Media: Kim Yi
Ebook Producer: Sophia Purut
Marketing Director, Chemistry: Stacy Loyal
Design Director: Rubina Yeh
Designer: Anne-Michelle Gallero
Director of College Permissions: Megan Schindel
College Permissions Specialist: Josh Garvin
Photo Editor: Mike Cullen
Composition: GW, Inc./Project Manager: Gary Clark
Illustrations: Alicia Elliott, Spark Life Science Visuals
Manufacturing: Transcontinental Interglobe—Beauceville, Québec

ISBN 978-1-324-04387-4 (paperback)

W. W. Norton & Company, Inc., 500 Fifth Avenue, New York, NY 10110
wwnorton.com

W. W. Norton & Company Ltd., 15 Carlisle Street, London W1D 3BS

To my family, who always inspire me

To my students, who invariably teach me

Brief Contents

Contents

Jeffrey B. Banke/Shutterstock.

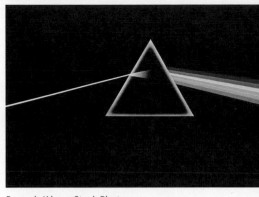

Records/Alamy Stock Photo.

Techniques Videos

Aqueous–Organic Extractions and Drying Organic Solutions

Boiling Point Measurement

Cleaning Glassware

Column Chromatography

Fractional Distillation

Gravity Filtration

Heating at Reflux

Melting Point Measurement

Neutralizing Acidic or Basic Solutions and Checking pH

Polarimetry

Preparing NMR Samples

Recrystallization

Rotary Evaporation

Simple Distillation

Thin-Layer Chromatography

Vacuum Filtration

Preface

A LAB CURRICULUM THAT EMPHASIZES CONNECTIONS AMONG CHEMISTRY, BIOLOGY, AND SUSTAINABILITY

Dear Instructor,

My mission in creating this textbook you're reading was to develop a green organic chemistry laboratory curriculum that inspires student engagement by emphasizing the connections among chemistry, biology, and sustainability. Incorporating concepts of green chemistry was a key aspect, not only to limit environmental impact and operational costs, but also to appeal to broader student and instructor interests. An important secondary consideration was to integrate the green chemistry with instruction in traditional techniques and glassware for the benefit of those students who need a practical foundation for careers in chemistry or other scientific disciplines and/or graduate study. Thirdly, the textbook needed to be flexible for use in one- or two-semester formats, whether the lab is integrated with the lecture or as a separate course.

How did this project start? At Iowa, we have a one-semester Organic Laboratory course for non-majors that meets twice a week plus lecture, with an enrollment of 150–200 per semester, including some chem majors who take it for scheduling reasons. Our non-majors course was taught without any major curriculum revision for many years. Over this time, many informal discussions about the need for revising some experiments, mainly for reasons of modernization and safety, prompted me to take action. Our graduate student teaching assistants and departmental lab staff recognized the need as well, and were willing and eager to test new experiments, so I began a significant curriculum revision, replacing a couple of experiments per semester. After a few semesters it became apparent that an emphasis on biological perspectives and sustainability had emerged as a coherent theme, and that it was an effective way to reach students with wide-ranging interests. It was then I realized that this emphasis, new for our course, could likely be of interest to the larger Chemistry Education community as well.

Why is a new book needed? Over the years, existing textbooks have tackled waste disposal, safety, and cost issues by downsizing the scale of reactions using specialized glassware that students won't likely see anywhere else. However, this approach is unsuitable for students who will need familiarity with standard glassware and realistic preparative scale reactions, whether it's in future employment or in graduate study. The microscale approach deals with the local classroom sustainability problem, but is less effective in teaching students to think about sustainability in real-world chemical processes where large scales are inevitable.

Compared with traditional lab texts—many of which contain hundreds of pages that are often not used in the typical undergraduate setting—a single text combining green chemistry principles with traditional organic techniques instruction offers cost and convenience advantages to both students and instructors. There is a clear need for this new textbook.

Goals

This textbook is available in two formats: with Techniques chapters standalone, or with Techniques and Experiments combined. My motivations for the combined textbook likely mirror concerns that inspire curriculum revisions at other institutions. I wanted to

a. replace certain time-worn experiments that had grown somewhat stale,

b. enhance connections to biological chemistry,

c. devote increased attention to issues of sustainability, and

d. better coordinate topics between lecture and lab.

At the same time, I recognized that many instructors have already assembled a set of experiments to address similar issues, and may wish to have a more cost-effective and focused Techniques component to accompany their experiments. That's why the two formats are available. In addition, Norton offers custom options that blend this book's chapters with your own experiments.

BRIEF OVERVIEW OF THE BOOK

When this textbook began to emerge from my course redesign, the potential advantages to a broader range of instructors and students became clearer. As seen in the table of contents, the organization of Techniques instruction places general topics (safety, recordkeeping, and report-writing) into Chapters 1–4, followed by standard separations, purifications, physical properties, and spectroscopy in Chapters 5–8. This Techniques volume allows for the instructor to reference specific chapters and sections as reading assignments, whether using the Techniques and Experiments combined textbook, or this Techniques textbook in conjunction with their own set of experiments.

One of the components of any instruction in green chemistry is minimizing the production of waste materials. At Iowa, our Environmental Health and Safety officials track the amount of waste generated by the Chemistry Department, and gratifyingly, with the inception of green chemistry experiments in our curriculum, waste quantities were reduced to two-thirds of the prior amounts. Now, when I discuss the 12 Principles of Green Chemistry with my students, I can point to our own class as an example of the tangible impact we can all have upon environmental protection. My hope is that this Techniques text will facilitate incorporating green chemistry experiments at your own institution, and will yield similar outcomes.

CONCLUDING REMARKS

During our development toward the goals outlined above, student engagement improved as judged by comments in student evaluations, and my own interest in teaching the course strengthened. What a great synergy!

Our non-majors' organic laboratory course at Iowa, with strong themes of biological chemistry and sustainability woven throughout, has proved very appealing to many hundreds of students who have already used the curriculum in the combined Techniques and Experiments version of this textbook. Their unsolicited positive comments and sincere engagement in the course have been very gratifying, and inspired my proposal to take the text to a wider audience. Quite a lot of critical thinking and adaptations based on instructor, student, teaching assistant, and reviewer feedback have been implemented to make this material more generalizable to other universities and colleges.

I thank you for taking the time to consider this textbook for your course, and I welcome your feedback so that I can continue to improve it in the years to come.

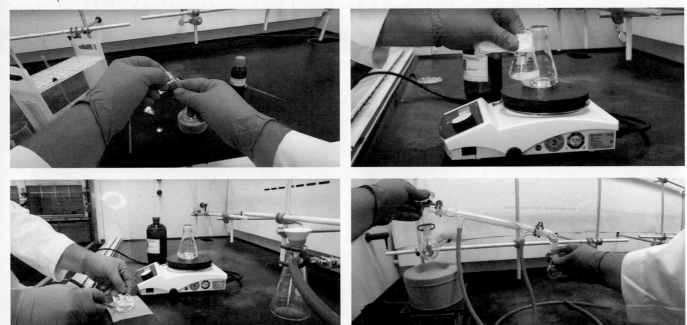

Courtesy of Gregory K. Friestad.

Developed by Dr. Friestad in collaboration with the Norton Chemistry team, 16 videos focus on the most common laboratory techniques used in the course. These videos are integrated into the ebook in both the Techniques chapters and the Experiments chapters where they are referenced, and are also included in the Smartwork pre-lab activities, ensuring students receive the support they need.

Preface for Students

A NOTE TO STUDENTS

Learning organic chemistry is an exciting endeavor because organic chemistry is the language of life. Indeed, organic chemistry is critical to communication, energy transfer and storage, nutrient uptake, growth, replication, and virtually everything else needed for life to exist on earth. Our understanding and manipulation of organic chemistry also impacts our daily lives in uncounted ways: treating our diseases (drugs), fueling our transportation (petroleum), maintaining our bodies (food), coloring our clothes (dyes), and constructing our homes (building materials). And don't forget, you wouldn't have mobile communication devices without organic chemistry. Understanding something so ubiquitous in our daily lives, and so critical to life itself, clearly enriches us.

Ask organic chemists what attracted them to the field, and most will recall their first opportunity to get into the lab, tinkering with the glassware, manipulating beautiful crystals or colorful liquids, and finding something new or unexpected. This laboratory course is intended to introduce you to those appealing joys, while also strengthening your understanding of the lecture material. The lab provides technical tools to be sure, but also builds practices of logical thinking that you'll find valuable in the future, wherever you are.

While many fantastic innovations have emerged from organic chemistry, we also know that hazards can exist throughout product life cycles, from design, development, production, and storage, then on through the use (or misuse) and disposal of the products. Lack of attention to potential hazards can create unintended consequences now or in the future. Can we anticipate these consequences, and minimize them, while we continue to creatively address the scientific questions and technological goals that are at the core of the human condition? I believe we can, and I want you and future students to learn organic chemistry while keeping this perspective in mind. That is part of the reason for this book.

Another important goal for me is to provide a cost-effective option for students in the organic chemistry lab. You likely have an organic chemistry textbook to accompany the lecture. Other lab textbooks can be just as large as that one, with hundreds of pages that go unused. I wanted to provide a more focused book, avoiding an "everything-under-the-sun" approach while delivering the instructional material you will need to succeed in the course, and beyond.

How to Use This Book

This book is your resource for instructions on how to carry out the common procedures that are required in the experiments your instructor assigns. A table of contents of this book, kept handy as you prepare for each experiment, will help you locate the background information you'll need to perform the various techniques that are found in the experimental procedures. Then, you can study or revisit any background reading before performing the techniques. This will help you understand how each procedure works and why it is needed in the experiment.

Additionally, pay attention to the Learning Objectives provided in each chapter. These indicate key points that you should be able to explain, or key things you should be able to do, after reading the chapter and performing the techniques in the course of your experiments.

This textbook also comes with 16 Techniques Videos that you can use throughout the course—first, to familiarize yourself with each technique, then to examine again when preparing to conduct a specific experiment. These videos can help you build confidence before entering the lab, by showing that technical skills are accessible to you, even on the first try.

The process of learning organic chemistry will benefit you in intangible ways, with a fun and appealing combination of both logic and creativity. This combination stimulates the development of valuable critical thinking skills, which will be useful no matter what career path you eventually choose. In the lab, you will add hands-on experience that deepens your understanding of all the knowledge and skills you learn in the lecture classroom. I hope that you'll find that this engaged learning experience will reveal the lively and adventurous nature of organic chemistry as an endless source of innovation and a strong foundation for the health sciences.

Good luck, be safe, and have fun!
Greg Friestad

Acknowledgments

REVIEWERS

Carolina Andrade *(Lone Star College)*
Michael Ansell *(Las Positas College)*
Cosimo Antonacci *(Seton Hall University)*
Jesse Bergkamp *(California State University—Bakersfield)*
Michelle Boucher *(Utica College)*
Laura Brown *(Indiana University)*
Kathleen Brunke *(Christopher Newport University)*
Bobby Burkes *(Grambling State University)*
Christopher Callam *(The Ohio State University)*
Chad Cooley *(Indiana University)*
Cliff Coss *(Northern Arizona University)*
Sean Curtis *(Des Moines Area Community College)*
Nathan Duncan *(Maryville College)*
Jason Dunham *(Ball State University)*
Brendan Dutmer *(Highland Community College)*
Ola El-Rashiedy *(Penn State University–Abington)*
Douglas Flournoy *(Indian Hills Community College)*
Nicholas Greco *(Western Connecticut State University)*
Dustin Gross *(Sam Houston State University)*
Matthew Grote *(Otterbein University)*
Scott Hartley *(Miami University)*
Nicholas Hill *(University of Wisconsin—Madison)*
Daniel Holley *(Columbus State University)*
Kevin Jantzi *(Valparaiso University)*
Keneshia Johnson *(Alabama A&M University)*
Michael Justik *(Penn State Erie, The Behrend College)*
Renat Khatmullin *(Middle Georgia State University)*
Mike Koscho *(University of Oregon)*
Joseph Kremer *(Alvernia University)*
Shane Lamos *(Saint Michael's College)*
F. Andrew (Andy) Landis *(Penn State University–York)*
Rita Majerle *(Hamline University)*
Rock Mancini *(Washington State University)*
Sara Mata *(Indiana University)*
Vanessa McCaffrey *(Albion College)*
Sri Kamesh Narasimhan *(SUNY Corning Community College)*
Erik Olson *(Upper Iowa University)*
Steve Oster *(Middlebury College)*
Hasan Palandoken *(California Polytechnic State University–San Luis Obispo)*
Noel Paul *(The Ohio State University)*
Angela Perkins *(University of Minnesota)*
Joanna Petridou-Fischer *(Spokane Falls Community College)*
Brian Provencher *(Merrimack College)*
Matt Siebert *(Missouri State University)*

Chester (Chet) Simocko *(San José State University)*

Mackay Steffensen *(Southern Utah University)*

Anne Szklarski *(King's College)*

Gidget Tay *(Pasadena City College)*

Matthew Tracey *(University of Pittsburgh)*

Michael Wentzel *(Augsburg University)*

Laura Wysocki *(Wabash College)*

Kimo Yap *(California State University–Los Angeles)*

Hui Zhu *(Georgia Tech)*

About the Author

Courtesy of Gregory K. Friestad.

GREGORY K. FRIESTAD is associate professor in the Department of Chemistry at the University of Iowa, with research interests in asymmetric synthesis, free radical chemistry, organosilicon chemistry, and new synthetic methods with transition metal reagents and catalysts. Dr. Friestad earned a BS degree in chemistry from Bradley University and a PhD in organic chemistry from the University of Oregon, mentored by Bruce P. Branchaud.

Dr. Friestad teaches at the graduate and undergraduate levels, including more than 15 semesters teaching majors and non-majors organic laboratory courses. Since 2015, Friestad has been developing a new curriculum for the introductory organic laboratory, focusing on standard-scale experimental procedures to develop skills and experience that translate to research, while emphasizing sustainability and biological perspectives. Aside from practicing organic chemistry, he enjoys travel, live music, disc golf, and spending time outdoors with family.

PART A
TECHNIQUES

1

LEARNING OBJECTIVES

- Relate the importance of organic chemistry to humanity, using the example of antibiotics.

- Recognize how green chemistry principles can minimize undesirable impacts of organic chemistry.

- Apply specific metrics of green chemistry to measure the impacts of organic chemistry processes or procedures.

Organic Reactions and the Twelve Principles of Green Chemistry

ORGANIC CHEMISTRY

Organic chemistry happens all around us, and within us. We can use organic chemistry to create new substances with new properties; this creativity can coexist with all of the fascinating organic chemistry in nature.
Courtesy of Gregory K. Friestad.

Chemistry has had a profound impact on human existence. The reactions of organic chemistry are how we convert readily available carbon-containing substances to new compounds and materials of greater value. This has fueled revolutionary innovations in all kinds of items in our daily experiences, from building materials to electronic devices, and from textiles to medicines such as the penicillins (**Figure 1.1**). Since organic reactions are involved in all these endeavors, learning the laboratory practice of organic chemistry can put you in a better position to understand these innovations, whether you intend to be on the front lines of discovery or a responsible end user of new inventions.

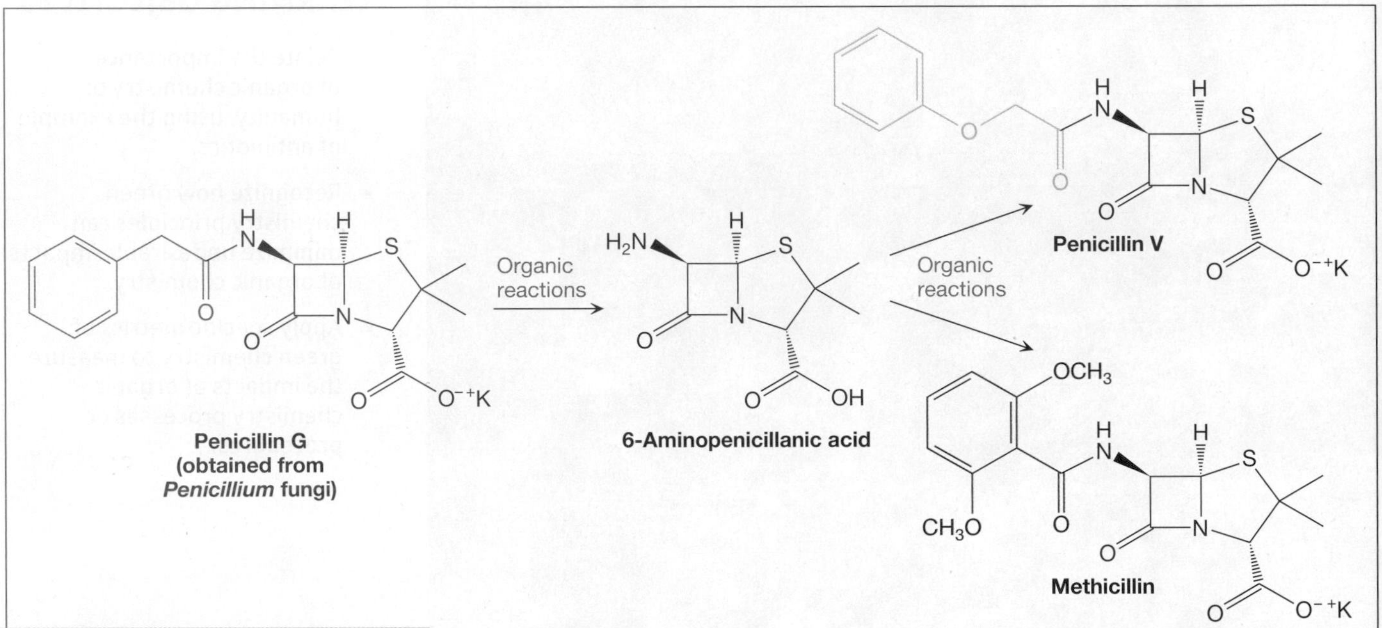

FIGURE 1.1

When bacteria developed resistance to penicillin G, transformations of the original compound to new penicillins, using organic reactions, led to the discovery and development of new antibiotics.

One of organic chemistry's most dramatic impacts has been on antibiotic drugs and the effects they have had on average life expectancy through the 20th century. Life expectancy worldwide increased dramatically from 1900 to 2010, and a comparison of the top 10 leading causes of death in 1900 versus 2010 indicates a connection to the widespread use of antibiotics (**Figure 1.2**).[1] The three top causes of death in 1900 were infectious diseases, for example, gastrointestinal infections. By 2010 these had almost disappeared from the top 10,[2] and during this same time period, there was a 30-year increase in life expectancy. In 1900, before antibiotics became a widespread tool for the treatment of infectious diseases, 30.4% of all deaths occurred among children ages 5 and below; in 1997, it was only 1.4%.[3]

A period of rapid increase in life expectancy came during 1936–1952, when antibiotics—first sulfa drugs, then penicillins—became widely introduced to the

[1]Arias, E. United States Life Tables, 2002. *Natl. Vital Stat. Rep.* **2004**, *53* (6), 1–40. https://www.cdc.gov /nchs/data/nvsr/nvsr53/nvsr53_06.pdf (accessed April 2022).
[2]Jones, D. S.; Podolsky, S. H.; Greene, J. A. The Burden of Disease and the Changing Task of Medicine. *N. Engl. J. Med.* **2012**, *366*, 2333–2338. DOI: 10.1056/NEJMp1113569
[3]U.S. Centers for Disease Control and Prevention. Achievements in Public Health, 1900–1999: Control of Infectious Diseases. *MMWR Morb. Mortal. Wkly. Rep.* **1999**, *48* (29), 621–629. https://cdc.gov/mmwr /preview/mmwrhtml/mm4829a1.htm (accessed April 2022).

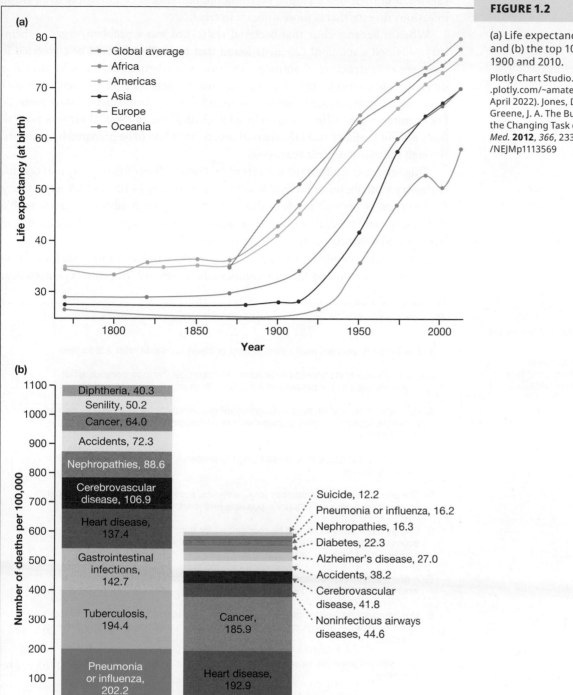

FIGURE 1.2

(a) Life expectancy change over time, and (b) the top 10 causes of death in 1900 and 2010.

Plotly Chart Studio. https://chart-studio .plotly.com/~amatelin/320 (accessed April 2022). Jones, D. S.; Podolsky, S. H.; Greene, J. A. The Burden of Disease and the Changing Task of Medicine. *N. Engl. J. Med.* **2012**, *366*, 2333–2338. DOI: 10.1056 /NEJMp1113569

public. Unfortunately, the industrial scale production and use of these compounds has also had unintended consequences in the form of antibiotic resistance.[4] Because microbial organisms rapidly reproduce, they evolve new characteristics with alarming ease. A few in the population may have chance genetic mutations that allow them to escape the effects of the antibiotic. These few resistant bacteria

[4]King, A. Why Antibiotic Pollution Is a Global Threat. *Chemistry World*. The Royal Society of Chemistry, 2018. https://chemistryworld.com/news/why-antibiotic-pollution-is-a-global-threat/3009021.article (accessed April 2022).

survive, and their descendants, carrying the mutations, can multiply to cause an infectious disease that is more difficult to treat.

When it became clear that bacterial resistance was a problem, organic chemistry provided a solution! Chemists found that penicillin G could be converted to a simpler core structure, 6-aminopenicillanic acid (Figure 1.1). Then, by attaching different side chains to the amino group, many variations on the penicillin core structure were invented, and some were found to be active against resistant bacteria. For example, methicillin was introduced to clinical use in 1960. These new penicillins were not available from the natural source; they had to be prepared by chemists, through the use of organic reactions.

This brief case study of the impacts of antibiotics illustrates one way that organic chemistry contributes to incredible advances in quality of life and life expectancy. However, every organic reaction that contributes to such advances comes with a cost: Energy and natural resources are consumed, waste is generated, accidents can happen, and products require end-of-use disposal.

Fortunately, chemists are in a position to understand and anticipate the costs of the chemistry we do and the new compounds we invent! Therefore, chemists bear

FIGURE 1.3

The originally published 12 principles of green chemistry.

Anastas, P. T.; Warner, J. C. *Green Chemistry: Theory and Practice*; Oxford University Press: New York, 1998.

1. It is better to prevent waste than to treat or clean up waste after it is formed.

2. Synthetic methods should be designed to maximize the incorporation of all materials used in the process into the final product.

3. Wherever practicable, synthetic methodologies should be designed to use and generate substances that possess little or no toxicity to human health and the environment.

4. Chemical products should be designed to preserve efficacy of function while reducing toxicity.

5. The use of auxiliary substances (e.g., solvents, separation agents, etc.) should be made unnecessary wherever possible and innocuous when used.

6. Energy requirements should be recognized for their environmental and economic impacts and should be minimized. Synthetic methods should be conducted at ambient temperature and pressure.

7. A raw material or feedstock should be renewable rather than depleting wherever technically and economically practicable.

8. Unnecessary derivatization (blocking groups, protection/deprotection, temporary modification of physical/chemical processes) should be avoided whenever possible.

9. Catalytic reagents (as selective as possible) are superior to stoichiometric reagents.

10. Chemical products should be designed so that at the end of their function they do not persist in the environment and break down into innocuous degradation products.

11. Analytical methodologies need to be further developed to allow for real-time, in-process monitoring and control prior to the formation of hazardous substances.

12. Substances and the form of a substance used in a chemical process should be chosen so as to minimize the potential for chemical accidents, including releases, explosions, and fires.

a responsibility to take a leadership role not only in providing the innovations that society demands, but also in minimizing their costs. Throughout this text, you will learn the fundamentals of laboratory organic chemistry while considering various ways to minimize your impact on health, safety, and the environment. Let's start with some guiding principles.

1.1 THE TWELVE PRINCIPLES OF GREEN CHEMISTRY

In the late 1990s, synthetic organic chemists began to look at preventing negative impacts of their endeavors through a new and broader perspective, called **green chemistry**, which is summarized by a set of principles popularized by Paul Anastas and John Warner.[5] The purpose of green chemistry is to minimize the risk of negative outcomes while enabling innovations and creativity to flourish. It recognizes the value and power of chemistry, while calling on chemists to prevent negative impacts of their work in advance and by design. This is a long-term problem that requires consistent effort and new ideas, but significant progress has been made and will continue. There is reason for great optimism, especially with new generations of chemists approaching the field with exposure to green chemistry and sustainability always in mind.

<< **green chemistry**
The design and implementation of products and processes that minimize or eliminate hazards of chemistry activities during all phases of product life cycles, including manufacture, use, and disposal.

The 12 principles of green chemistry are presented in their originally published form in **Figure 1.3**. The sections that follow present some examples of how these pertain to organic reactions and the organic chemistry laboratory studies upon which you are about to embark.

1.1A Prevent Waste: Avoid Producing Waste, so There Is No Need for Treatment or Cleanup

Certain classes of reactions inherently produce more waste than others. Among the common reaction classifications (**Figure 1.4**), substitutions and eliminations by their very nature will produce one molar equivalent of waste for the amount of product formed. Even if the reaction can be improved to 100% efficiency, significant quantities of waste are formed. A useful metric for evaluating this is the environmental impact factor or **E-factor**,[6] which considers not only the yield of product, but also all other outputs of waste materials. It's worthwhile to do this calculation for experiments that you perform. For an aqueous solution of waste, consider only the mass of the waste solute; water is generally omitted from the E-factor calculations. In the companion Experiments manual, a number of reactions have been designed for aqueous conditions (or aqueous ethanol), thereby substituting innocuous and renewable solvents in order to minimize the output of flammable, petroleum-sourced, or halogenated solvents. An ideal synthesis would have an E-factor of zero. This calculation can use kilograms or grams, as long as the units match in the numerator and denominator.

<< **E-factor**
A measure of environmental impact of a chemical process, defined as the mass of waste produced divided by the mass of product obtained.

$$\text{E-factor} = \frac{\text{mass of waste (kg)}}{\text{mass of product (kg)}}$$

[5]Anastas, P. T.; Warner, J. C. *Green Chemistry: Theory and Practice*; Oxford University Press: New York, 1998.
[6]Sheldon, R. A. Organic Synthesis: Past, Present and Future. *Chem. Ind.* **1992**, *23*, 903–906.

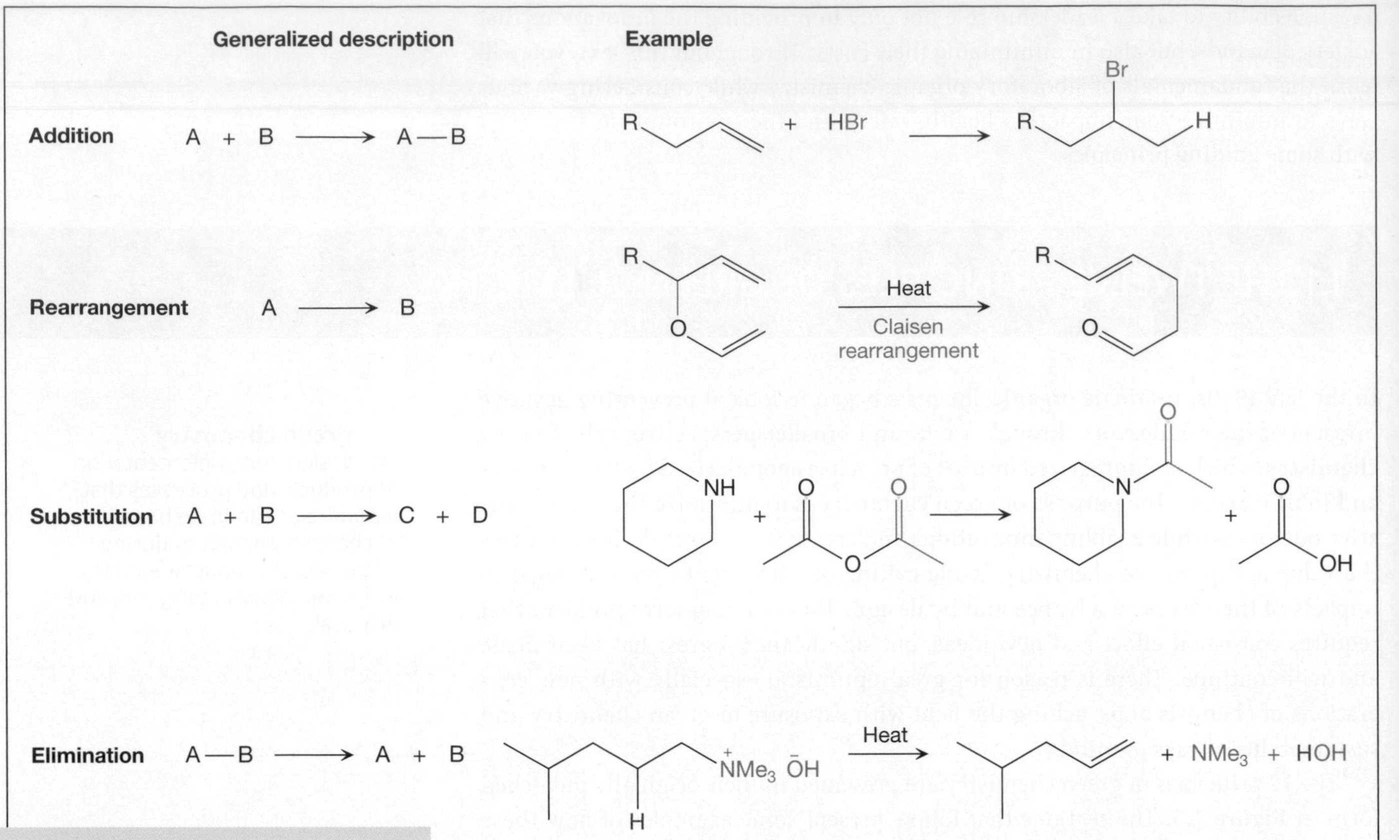
Generalized description
Example
Addition
A + B ⟶ A—B
R
+ HBr
Br
R
H
Rearrangement
A ⟶ B
R
O
Heat
Claisen
rearrangement
R
O
Substitution
A + B ⟶ C + D
NH
+
O
O
O
N
O
+
O
OH
Elimination
A—B ⟶ A + B
H
+
NMe3 OH
Heat
+ NMe3 + HOH

FIGURE 1.4

Reaction types and examples illustrating side products and atom economy.

1.1B Atom Economy: Incorporate All Atoms from Starting Materials into Products

Atom economy is a measure of reaction efficiency that considers the outcome for all the atoms of the various reaction inputs. Certain types of reactions are very *high* on atom economy, such as addition reactions (Figure 1.4), because two reactants become one. If, on the other hand, most of the atoms from a very large reagent end up as by-product waste material, then the process has very *low* atom economy. In the companion Experiments manual, there are some cycloaddition reactions that incorporate all of the reactants into the products for excellent atom economy.

Atom economy is calculated in two ways, called intrinsic and experimental. Intrinsic atom economy can be calculated based on theory, from the balanced equation on paper, whereas experimental atom economy uses data from laboratory results.[7] Experimental atom economy is a somewhat confusingly named quantity because it uses **theoretical yield** as the numerator. The name comes from the denominator, where any excess amounts of reagents that are used in the lab are included, even if their amounts are beyond the stoichiometry required by the balanced reaction equation.

theoretical yield >>
The maximum amount of product that could be obtained if all of the limiting reagent is converted to the product.

$$\text{Intrinsic atom economy (\%)} = \frac{\text{molar mass of desired product}}{\text{molar mass of all reactants}} \times 100$$

$$\text{Experimental atom economy (\%)} = \frac{\text{theoretical yield of desired product (g)}}{\text{actual quantity of all reactants used (g)}} \times 100$$

[7](a) Trost, B. M. The Atom Economy—A Search for Synthetic Efficiency. *Science* **1991**, *254*, 1471–1477.
(b) For several examples of atom economy calculations, see Abhyankar, S. B. Introduction to Teaching Green Organic Chemistry. In *Green Organic Chemistry in Lecture and Laboratory*; Dicks, A.P., Ed.; CRC Press: Boca Raton, FL, 2012; pp 1–28.

Worked Example

Calculate the E-factor, intrinsic atom economy, and experimental atom economy for the S_N2-type Williamson ether synthesis from 2-naphthol and 1-iodobutane (**Figure 1.5**).

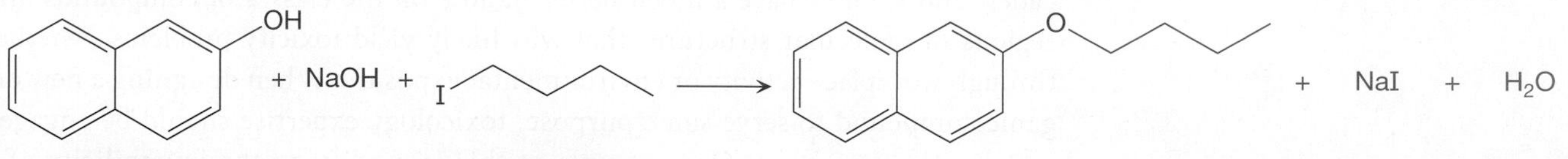

2-Naphthol

1.00 g
(144.2 g/mol)
Limiting reagent

0.560 g
(40.0 g/mol)

1-Iodobutane

1.60 g
(184.0 g/mol)

1.29 g
(200.3 g/mol)

(149.9 g/mol)   (18.0 g/mol)

$$\text{Theoretical yield (mol)} = \frac{1.00 \text{ g}}{144.2 \text{ g/mol}} = 0.00693 \text{ mol}$$

$$\text{Experimental atom economy} = \frac{0.00693 \text{ mol} \times 200.3 \text{ g/mol}}{3.16 \text{ g}} \times 100\% = \boxed{44\%}$$

FIGURE 1.5

The S_N2 reaction between 2-naphthol and 1-iodobutane, with experimental amounts of the reactants and the main organic product.

To calculate the E-factor, we need to know the mass of waste material ($NaI + H_2O$) and the mass of desired product (the ether). The combined mass of NaI and H_2O can be calculated by subtracting the mass of desired product (1.29 g) from the sum of all the reactant masses (1.00 + 0.560 + 1.60 = 3.16 g). The remainder (3.16 – 1.29 = 1.87 g) is the amount of waste. The mass of waste divided by the mass of product is the E-factor.

$$\text{E-factor} = 1.87 \text{ g}/1.29 \text{ g} = 1.45$$

For the intrinsic atom economy, we first divide the molecular weight of the desired product (200.3 g/mol) by the sum of molecular weights of reactants, with each reactant molecular weight multiplied by the stoichiometry of the reactant in the balanced chemical equation. We then convert that value to a percentage by multiplying by 100%. In this case, the stoichiometry is 1:1:1, which simplifies the calculation:

$$\text{Intrinsic atom economy} = 200.3/(144.2 + 40.0 + 184.0) \times 100\% = 54\%$$

The experimental atom economy is the theoretical yield of the ether in grams, divided by the mass in grams of all of the reactants used, and which we calculated previously to be 3.16 g. As shown in Figure 1.5, the theoretical yield in moles is the amount of moles of the limiting reactant, 2-naphthol, which is obtained by dividing the amount used by its molecular weight. Multiplying this by the molecular weight of product gives a theoretical yield in grams of 1.39 g. We then divide this value by the sum of the reactant masses used and convert to a percentage:

$$\text{Experimental atom economy} = (1.39 \text{ g})(3.16 \text{ g}) \times 100\% = 44\%$$

Notice that this experimental atom economy is lower than the intrinsic atom economy. This is because excess amounts of NaOH and 1-iodobutane were used, beyond the 1:1:1 stoichiometry of the balanced reaction equation.

1.1C Low Toxicity: Design Alternative Processes/Substances That Are Known to Have Lower Toxicity

The field of toxicology has advanced by leaps and bounds over the last few decades, and we now have a much better handle on the classes of compounds and regions of molecular structures that will likely yield toxicity problems, whether through workplace activity or environmental exposure. When designing a new organic compound to serve some purpose, toxicology expertise should be engaged early in the discussion. One example of this is modifying the lipophilicity of a compound, which is known to affect the degree to which it is transported across membrane barriers at the cellular level or the skin. If altering lipophilicity can minimize transport of a compound through the skin, then its exposure hazards can be reduced, and a less toxic material could be designed with the same intended function. Well-informed decisions up front may avoid wasting additional resources by reversing course later.

1.1D Maintain Function While Lowering Toxicity

Achieving a greater ratio of effectiveness to toxicity is important, whether it is by lowering the toxicity or by increasing the efficacy. Either approach is beneficial. A more highly efficacious compound may be used in lower quantity, so the by-products of its manufacture and use will also be minimized. The key point here is that risk is a function of both hazard and exposure: If a substance retains its function even when used in very low quantities, then its risk is lowered simply because the amount of the exposure is lower.

1.1E Avoid Solvents or Separation Agents Whenever Possible

column chromatography >>
A purification technique whereby a mixture of organic products is separated into its components due to differences in the rate of travel with a solvent through a column of insoluble powdered material, generally silica gel.

In an organic synthesis, a common purification method is **column chromatography**, which consumes large quantities of solvents and adsorbents. Sometimes these can be recycled, but sometimes they add to the waste stream without creating any new bonds or structural changes in the compound. On the other hand, if reactions can be designed so that the product can be moved on to the next reaction step without purification, all of this waste can be avoided.

1.1F Energy Efficiency: Minimize Impacts by Working at Ambient Temperature and Pressure

It takes a lot of energy to perform reactions outside of ambient conditions. Heating requires an input of energy, but cooling is energy intensive, too, because electricity is needed to run refrigerator devices and produce coolants like dry ice or liquid nitrogen. Pressurized reactions, whether high pressure or low pressure, require pumps that also use energy. Removing solvent requires energy input, too. Bristol Myers Squibb won a Presidential Green Chemistry Challenge Award for its redesign of the synthesis of the cancer drug Taxol. The redesigned route eliminated 10 solvents and improved energy efficiency by avoiding six drying steps.[8] On the much smaller academic lab scale, these factors may seem negligible. But, if reactions can be designed to avoid these energy inputs, then the energy efficiency improvements become very significant upon scale-up to industrial production.

1.1G Renewable Feedstocks: Avoid Depleting Natural Resources When Feasible

Organic chemistry makes extensive use of petroleum feedstocks, so organic chemists are often dismayed to see all these valuable materials simply burned away in combustion engines. We know that petroleum feedstocks are a finite resource, and gradually people are finding ways to access alternative feedstocks from agricultural products or waste materials. A well-publicized example is the use of corn and its by-products to generate ethanol. But beyond their use as fuel, a wider variety of chemical feedstocks can be made accessible as organic chemists develop ways to convert waste into replacements for petroleum products. One experiment in the companion Experiments manual shows you how to convert corncobs into furfural, which is the starting material for an addition reaction to form a carbon–carbon bond.

1.1H Fewer Synthetic Steps: Avoid Unnecessary Derivatization (e.g., Protecting Groups)

Synthesis sequences made up of multiple steps are prone to problems of functional group incompatibility along the way. In a hypothetical sequence of six steps, the functional group formed in step 3 may not be compatible with the reaction planned for step 6. To get around this, an extra reaction may be needed to protect the vulnerable functional group from an undesired reaction. Such extra steps can be detrimental to the overall efficiency of the synthesis sequence, and if possible, the sequence should be redesigned to avoid extra steps.

1.1I Catalytic Processes: Use Catalytic Processes of Superior Efficiency Relative to Stoichiometric Ones

Catalysis has remarkable potential to improve the efficiency of synthesis. Instead of using equal molar amounts of a reagent that ends up producing equivalent amounts of waste, a catalytic process recycles the reagent during the reaction. In this way, each molecule of the reagent is reused repeatedly (sometimes thousands of times). Nature uses enzymes in this way. Imagine if you had to consume equal amounts of reagents to digest all the food you eat. Instead, your digestive tract contains very small quantities of enzymes to do this work. Because they are catalytic, the enzymes are reused over and over, so they contribute almost nothing to the waste output. Furthermore, they lower the energy of activation of reactions, allowing them to occur at lower temperatures, minimizing energy inputs. In the companion Experiments manual, two experiments use a tungsten **catalyst** for oxidations, and

<< **catalyst**
A compound that lowers the energy of activation for a chemical reaction but is not consumed by the reaction.

---

[8]U.S. Environmental Protection Agency, Office of Pollution Prevention and Toxics. Presidential Green Chemistry Challenge Award Recipients 1996–2016, 2016. https://epa.gov/greenchemistry/document -green-chemistry-challenge-award-recipients-1996-2016 (accessed April 2022).

in two other experiments, biocatalysis (vitamin B_1 and a ketoreductase enzyme) is used for the nucleophilic addition to a carbonyl and the **enantioselective** reduction of a ketone, respectively.

1.1J Innocuous After Use: Avoid Products That Persist After Use and Are Unsafe After Degradation

There are numerous heartbreaking stories of unanticipated outcomes from industrial wastes that were not disposed of by today's standards, and continue to persist in the environment or accumulate in organisms. In Times Beach, Missouri, for example, which is a town outside of St. Louis, an industrial waste oil containing dioxin was sprayed on roads in an effort to keep dust down. It persisted in the environment and reached a level from 300 to 700 times greater than that deemed safe by the U.S. Centers for Disease Control and Prevention. Eventually the entire town was purchased by the federal government and the residents resettled elsewhere to avoid further exposures. Awareness raised in response to events such as these has led to much better standards for waste handling, but accidents still happen. If an alternative to dioxin could have been designed that degrades on exposure to light, this problem may have been avoided. Situations such as these present opportunities and responsibilities for organic chemistry, and designing compounds that degrade in a harmless fashion is an active area of research.

1.1K In-Process Monitoring: Control Processes in Real Time, Prior to the Formation of Hazardous Materials

A variety of negative impacts occur when a reaction gets out of control. For example, pressure may build up, reaction vessels may exceed their capacity, and accidents can result. Many of these situations can be addressed easily in an academic lab. On the industrial scale, however, these breakdowns can be much more significant. If the quality of the product is low because of a problem with process control, it may be unsuitable for the market and will become part of the waste stream. Or, a problem with temperature in a reactor may cause a nontoxic reaction to begin producing an unanticipated toxic by-product. These situations call for analysis of the reactions and processes throughout, so that any loss of control can be detected immediately and corrected.

1.1L Avoid Accident-Prone Materials: Avoid the Possibility of Release, Explosion, and Fire

Accidents in the lab are an important topic for everyone to discuss, so that they can be prevented when possible and so that their impact can be minimized when the unanticipated does occur. One way to do this is to avoid a reaction if you know that it comes with a high risk of release, explosion, or fire. The severity of the problems is magnified on a larger scale, so organic chemists should implement safer alternatives early in the development phase, before scaling up the synthesis for production.

Two different syntheses of oxirane (ethylene oxide) are shown in **Figure 1.6**. The first uses stoichiometric amounts of reagents to reach oxirane in two steps. The second is a one-step process using a catalyst (recall that a catalyst is recycled, not consumed, in a reaction). Refer to these syntheses to address the questions that follow.

1. (a) Prior to scaling up for industrial production, chemists developing the lab-scale synthesis of oxirane in Figure 1.6a found that a reaction producing 0.264 kg oxirane also produced 0.99 kg $CaCl_2$ and 0.33 kg HCl. Calculate the E-factor for this reaction under these conditions. (b) These chemists found that the synthesis in Figure 1.6b yielded 2.40 kg oxirane, with 0.72 kg waste material. Calculate the E-factor under these conditions.

2. Calculate the intrinsic atom economy for each of the syntheses in Figure 1.6.

3. Which of the syntheses in Figure 1.6 would be preferred from the green chemistry perspective? Use both E-factor and intrinsic atom economy to justify your answer.

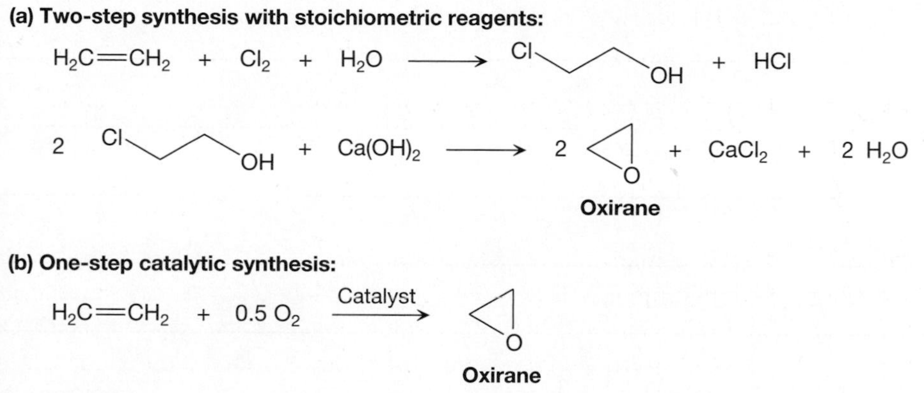

FIGURE 1.6

Two different routes for the industrial production of oxirane.

(a) Two-step synthesis with stoichiometric reagents:

(b) One-step catalytic synthesis:

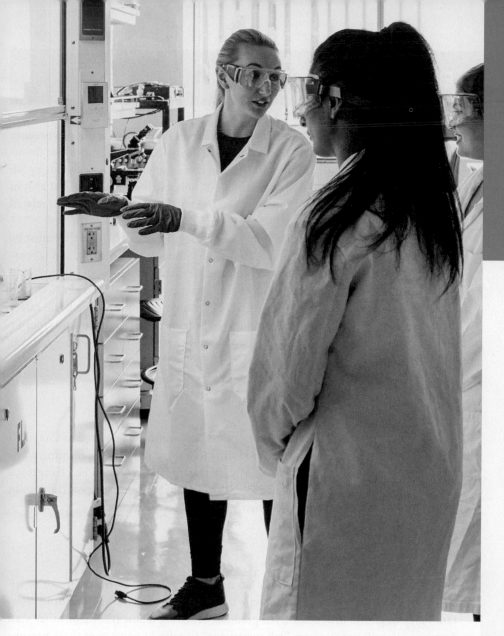

2

LEARNING OBJECTIVES

- Identify and implement safe practices in the organic chemistry laboratory.

- Locate and interpret hazard data using SDS information and standardized labeling symbols.

- Use appropriate personal protective equipment and approved waste disposal methods.

- Demonstrate knowledge of your location's safety procedures and equipment by taking a quiz.

Working Safely in the Organic Chemistry Laboratory

LAB SCENE

Workers have a responsibility to themselves and to others to work safely in the organic chemistry laboratory.
Thomas Barwick/Getty Images.

Instructional laboratories generally involve experiments that are chosen because they are relatively safe. Still, you need to learn the tools and procedures that chemists use to operate safely in the lab. This is a key component of any chemistry lab course, to keep everyone as safe as possible not only during the course, but also in future lab work. The skills you learn in this course will be useful in a variety of related fields.

Safety in the chemistry instructional laboratory is a responsibility that must be shared among all who will be present in the lab. Students, instructors, and laboratory staff all have the same interest in avoiding accidents. The topic of safety should be an ongoing discussion involving all of these participants, with updates to the lab procedures and guidelines as best practices continue to evolve. Some basic safety guidelines are a great starting point. Additional site-specific safety protocols may be slightly different than the general guidelines here, so be sure to follow the protocols provided by your lab instructor.

2.1A Preparation

Come to lab prepared. General instructions for preparation are covered in Chapter 3 of this text. Most of the risks associated with laboratory work can be minimized by thorough preparation prior to arrival. Any chemistry laboratory can be a dangerous place if procedures are not followed according to plan. You have to know the plan in order to follow it, so prepare in advance.

2.1B Supervision

Depending on your institution, your lab may be supervised by the professor or instructor in charge of the course, or your lab may be supervised by a teaching assistant (TA), associate instructor (AI), graduate teaching fellow (GTF), or other staff member. We will use the term "lab instructor" to refer to any one of these persons who may be supervising your lab.

Students are allowed in the laboratory only during their assigned times and with proper supervision. Do *not* enter the lab if one of these experienced authorities is not present. If you must leave the lab for any reason, inform your lab instructor.

2.1C Unauthorized Experiments

All experiments must be approved by the lab instructor. Instructional labs are not the time or place to invent new lab procedures or try out unauthorized experiments.

2.1D Laboratory Attire

Dress appropriately in the lab to minimize the risk of injury. Follow the attire guidelines in place at your school, which may be slightly different from these general guidelines:

- Wear shoes that cover the entire foot. Open-toed shoes, sandals, flip-flops, canvas shoes, or shoes with perforations are unacceptable.

- Cover your legs. Shorts, short skirts, and short dresses are unacceptable.
- Wear a shirt that covers you completely. Muscle shirts, tank tops, or anything that leaves your arms or torso exposed are unacceptable.
- Do not wear loose clothing.
- Tie back long hair.

2.1E | Hazard Data and SDS Access

Specific hazard data on millions of different compounds can be found in **Safety Data Sheets (SDSs)**. These are readily available online. For example, if you will be handling dichloromethane, you should first access the SDS to find out what hazards are listed for that compound. To find it, simply search "SDS dichloromethane" in your web browser. Chemical suppliers are required to provide the SDS for any product they sell. Keep in mind that the information you find there may be addressed to people who handle much larger quantities than we would use in an instructional lab.

<< **Safety Data Sheets (SDSs)** Safety advisory resources generally provided by chemical suppliers, providing hazard data for any chemical compound that is sold commercially. Also known as Material Safety Data Sheets (MSDSs).

2.1F | Hazard Labeling Pictograms

The Occupational Safety and Health Administration (OSHA) has standardized pictograms to alert users of various hazards presented by compounds, as shown in **Figure 2.1**. Always check the label of a chemical container to review these hazards before handling the chemical.

2.1G | Contact Lenses

Many schools and workplaces do not allow use of contact lenses in the laboratory without special approval due to medical reasons. If contact lenses are allowed at your location, they must be accompanied by goggles that completely contact the skin around the eyes to protect from splashes. Contact lenses are *not* eye protection devices.

2.1H | Food

Eating, drinking, or the use of any tobacco product (including e-cigarettes or vaping devices) is prohibited in the laboratory. This includes chewing gum, cough drops, throat lozenges, and the like.

2.1I | Medical Conditions

People with conditions that could be adversely impacted by exposure to organic chemicals should consult with their health care provider. Some organic chemicals are potential hazards specifically to the fetus or to young children. Those who are pregnant, nursing, or who suspect they may be pregnant are strongly advised to consider the advice of their health care provider, and may wish to take this course at a later time.

2.1J | Service Animals

Those with service animals should work in consultation with the lab instructor to determine how to safely comply with institutional policy guidance regarding service animals in labs.

FIGURE 2.1

OSHA.

Hazard Communication Standard Pictogram

The Hazard Communication Standard (HCS) requires pictograms on labels to alert users of the chemical hazards to which they may be exposed. Each pictogram consists of a symbol on a white background framed within a red border and represents a distinct hazard(s). The pictogram on the label is determined by the chemical hazard classification.

HCS Pictograms and Hazards

Health Hazard	Flame	Exclamation Mark
• Carcinogen • Mutagenicity • Reproductive Toxicity • Respiratory Sensitizer • Target Organ Toxicity • Aspiration Toxicity	• Flammables • Pyrophorics • Self-Heating • Emits Flammable Gas • Self-Reactives • Organic Peroxides	• Irritant (skin and eye) • Skin Sensitizer • Acute Toxicity (harmful) • Narcotic Effects • Respiratory Tract Irritant • Hazardous to Ozone Layer (Non-Mandatory)
Gas Cylinder	Corrosion	Exploding Bomb
• Gases Under Pressure	• Skin Corrosion/ Burns • Eye Damage • Corrosive to Metals	• Explosives • Self-Reactives • Organic Peroxides
Flame Over Circle	Environment (Non-Mandatory)	Skull and Crossbones
• Oxidizers	• Aquatic Toxicity	• Acute Toxicity (fatal or toxic)

For more information:

U.S. Department of Labor   www.osha.gov   (800) 321-OSHA (6742)

OSHA 3491-01R 2016

2.2 LABORATORY SAFETY EQUIPMENT

2.2A Fume Hood

Most organic chemistry operations should be conducted in a fume hood (**Figure 2.2a**). Fume hoods are cabinets with an exhaust flow designed to pull vapors out of the room from behind the work area, minimizing chemical exposure to people during lab activities. They generally have a hood sash with shatter-resistant glass that can be raised and lowered, and some also have panels that slide from side to side. As much as possible, chemists should work with the hood sash closed. Items that are not actively in use should not be stored in the hood; anything that inhibits the designed airflow can result in unnecessary hazard exposure. Hoods are not storage areas.

(a)

(b)

(c)

FIGURE 2.2

(a) A fume hood with shatter-resistant glass sash in the closed position. (b) An eyewash station. (c) A safety shower.

Courtesy of Gregory K. Friestad.

2.2B Personal Protective Equipment (PPE)

A variety of measures can be taken for personal protection during chemistry laboratory work. The type of PPE a chemist chooses will depend on the quantities of materials being handled and the types of hazards that will be encountered. In the undergraduate organic chemistry laboratory, safety goggles and gloves are generally suitable. If further measures are warranted, your lab instructor will notify you.

SAFETY GLASSES

You must wear eye protection at all times in the lab! Goggles that contact the skin in a continuous loop around the eye area are preferred, because they provide better protection from these hazards. The minimum eye protection should be safety glasses with side-shields to protect against splashes or objects that may approach from various directions.

GLOVES

Gloves may or may not be required for all experiments at your school. Make sure you know which experiments require gloves before you begin. Gloves may be reuseable or disposable. Reuseable gloves are generally thicker and more durable, and may be suitable for longer exposures. Disposable gloves, such as nitrile examination gloves, protect your hands from brief chemical exposures. They are not generally suitable protection for full immersion into chemicals other than water. If disposable gloves have holes, tears, discoloration, or swelling of the glove material after contact with a chemical, remove the gloves, wash your hands, and get a new pair. Be cautious about what you touch with gloves; if you touch pencils, pens, or cell phones, they'll be contaminated with whatever is on the outside of your gloves. *Gloves should not be worn outside the lab.* Always remove gloves before touching a doorknob or any other surface that others generally touch without gloves. If you see someone wearing gloves in the hall, how do you know which doors or other surfaces they may have contaminated?

2.2C Eyewash Station and Safety Shower

These are emergency tools that can prevent serious injuries in case of an accident. Know where eyewash stations and safety showers (**Figure 2.2b** and **Figure 2.2c**) are located in advance of the need to use them. If a chemical is splashed into the eyes, you do not want to waste any time trying to locate the eyewash. The safety shower, for emergency use only, will rapidly deposit a large amount of water, and is effective in the event clothing catches fire, or in case of a larger chemical spill affecting enough body area that it cannot be washed effectively at the sink.

2.2D Fire Extinguishers

Know the location of every fire extinguisher in the lab. Most organic solvents are flammable and ignite readily when exposed to a source of ignition. Thus, open flames are *not* permitted in the laboratory, unless specifically directed by your lab instructor. Smoking is strictly forbidden. In case of fire, it is also important to know the locations of safety showers and the nearest building exit.

2.3 WASTE DISPOSAL

Laboratory waste (solvents, solids, sharps, etc.) must be disposed of properly. **Table 2.1** lists general guidelines for waste disposal. **Figure 2.3** shows examples of waste containers that are properly labeled and stored. The labeling of containers and guidelines for disposal at your school may be somewhat different, and should be reviewed by your lab instructor before you begin the experiment. Make sure you know what type of waste you have before you dispose of anything. If you are unsure how to dispose of something, ask your lab instructor. No waste should ever go into the sink without permission.

TABLE 2.1

Waste Disposal Guidelines

CATEGORY	EXAMPLES	METHOD OF DISPOSAL
Organic waste	Waste solvents, acetone, reaction intermediates, products	"Organic Waste" bottles in hood, separate bottles for non-halogenated or halogenated (e.g., dichloromethane)
Aqueous waste, hazardous	Dilute acids and bases, aqueous washes from extractions	"Aqueous Waste" bottle in hood
Aqueous waste, nonhazardous	Water baths, ice baths, etc., that have *not* been contaminated with hazardous materials	Sink
Sharps	Broken glassware, test tubes, TLC plates, thermometers, etc., but *not* mercury thermometers!	Plastic "Sharps" bucket
Solid chemical waste, hazardous	Used filter papers, used drying agents, insoluble organic and inorganic solids	"Solid Waste" container in hood
Other solid waste, hazardous	Paper towels used for lab cleanup, used gloves, empty vials, other laboratory waste	Plastic solid waste bucket
Nonhazardous waste	Paper towels used for drying clean water from hands or glassware, other garbage not associated with laboratory activities	Trash container
Mercury	Broken thermometers or manometers	For any type of mercury spill, special cleanup procedures are required. Do *not* put mercury in the trash or sink. Contact laboratory staff immediately.

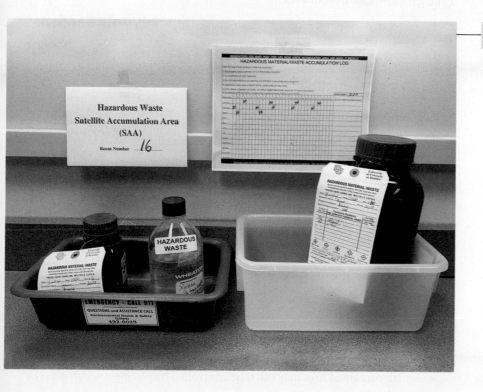

FIGURE 2.3

Waste accumulation site, with containers properly labeled. Note that liquid waste is kept within secondary containment (the plastic bins).

Courtesy of Derek J. Hayes, University of Colorado Boulder.

If an accident occurs, always inform your lab instructor and other laboratory staff as soon as it is safe to do so. If any accident involves a hazard that is beyond your capability to respond, remove yourself and others from the area immediately and notify your lab instructor and other laboratory staff.

2.4A Chemical Spills

Any spill should be cleaned up immediately using agents found in spill kits that are supplied in the lab room. These contain solid neutralizing agents that are designed for specific types of spills. Do not use the spill kit reagents for spills on your body (see Chemical Exposures, below).

- Acids are treated with a magnesium oxide–containing agent (e.g., Spill-X-A).
- Base spills are treated with a citric acid–containing agent (e.g., Spill-X-C).
- Solvent spills are treated with an absorbent solid agent (e.g., Spill-X-S).
- Use caution in responding to any liquid spill; many liquids look like water. Even if it is confirmed to be water, clean it up to avoid the hazards of a slip or fall.
- Report the use of spill cleanup reagents to your lab instructor, so the reagents can be replaced.

2.4B Chemical Exposures

Eyewash stations and safety showers are standard in organic chemistry labs. Your instructor is responsible for showing you the locations of these, and how to operate them.

If you get chemicals in your eyes,

1. go to an eyewash station and flush immediately with lots of water;

2. report the accident; and

3. get medical attention!

If you get chemicals on your skin,

1. go to the sink or safety shower and wash immediately with large amounts of water;

2. remove contaminated clothing;

3. continue to wash the area with water;

4. report the accident; and

5. get medical attention!

2.4C Fires

If there is a chemical fire in the hood or bench area:

- Small, contained fires may often be controlled by placing a watch glass or large beaker over the vessel.

- Larger fires may require the use of a fire extinguisher.
- Uncontrolled fires require using the alarm to contact the fire department.

If there is a fire on you or your neighbor,

- don't panic;
- shout for help;
- roll on the floor to smother the flames; and
- *walk* to the nearest safety shower.

2.4D | Broken Glassware

Cuts sustained from breaking glassware are among the most common lab injuries. Handle broken glassware with caution. All broken, cracked, or chipped glass should be disposed of in a properly labeled glass waste container.

If you break something, keep in mind what chemicals are in or on it. Alert your lab instructor and don't try to pick up small pieces with your fingers. Use a broom, dustpan, or other necessary cleanup supplies that are available, and avoid incidental skin contact.

2.4E | Injury

Report any injury to your lab instructor, no matter how small. For treatment of cuts, it is best to be on the safe side and visit a health care provider in case small bits of glass remain in the cut. Take note of what chemicals may have been present, in case medical personnel ask for this information.

Most schools require an incident report to be filed for any and all injuries, regardless of the severity. Use the reporting protocols as advised by your lab instructor.

2.5 SAFETY QUIZ INFORMATION

All students enrolled in chemistry lab courses should complete a safety quiz at the beginning of each course. The details of how this is administered may vary depending on your lab instructor, so check your syllabus or any other online course materials before the course begins. Many questions about safety can be answered using simple common sense, but you will also need to prepare by reading this chapter and any other safety materials assigned by your lab instructor. You should be able to answer some of the questions based on information discussed in your previous chemistry courses. Other questions may require some knowledge of the layout of the laboratory and building.

2.5A | Some Reminders to Assist You with the Quiz

- Safety goggles are the prime protection for your eyes. Contact lenses are *not* eye protection. In fact, contact lenses can trap liquids against your eye, making it more difficult to flush them out.
- Safety equipment, such as showers, eyewash stations, and fire extinguishers, are found in every lab. Be sure you know where they are in your lab. Have your lab instructor explain how to operate them.

- If you spill anything on your skin, wash it off immediately. Don't experiment by trying to run reactions on yourself or another student.
- If you or another student have an accident in lab, tell your lab instructor as soon as possible. Don't try to wait out the entire lab session.
- If you break glassware, don't just throw the broken pieces into any waste container. Check in your lab for a place where broken glass can be safely disposed of.
- Eating, drinking, and chewing gum are all prohibited in all chemistry labs, as is the use of all tobacco and vaping products.
- Closed-toed shoes (not sandals), long pants or skirts (not shorts or mini skirts), and shirts (not tank tops or muscle shirts) must be worn in the lab.
- Keep books, backpacks, and coats out of the aisles and off the bench tops. Ask your lab instructor where these items can be safely stored during lab periods.
- Check the location of the closest exits from your lab and from the building so that you can exit the building quickly in case of a fire.
- Never work alone or without supervision in the laboratory. Your lab instructor will be present throughout the entire lab period.
- Come to the lab prepared. Always be sure that you understand what you are doing before you do it. If you have questions, *ask*!

LEARNING OBJECTIVES

- Identify how records of daily lab activities affect interpretation, attribution, and reproducibility of scientific discoveries.

- Implement an orderly procedure for recording information and observations before, during, and after lab activities.

- Locate physical properties and hazard data for organic compounds.

- Create a flowchart to organize lab activities in advance.

Pre-Lab Preparation: The Laboratory Notebook

LIBRARY AND INTERNET SOURCES

Organic laboratory work requires advance preparation and information gathering, including searches of handbooks, databases, and the like.

Ammentorp Photography/Alamy Stock Photo.

INTRODUCTION

The laboratory notebook serves as a permanent chronological record of experimental work, whether it is prepared in hardcopy format or with digital notebook software. When there is a legal dispute over who was the first to find a lucrative new discovery or invention, the laboratory notebook is a key piece of evidence. So, keeping a proper lab notebook could be worth a lot of money and prestige! Also, sometimes you need to reproduce an experiment exactly as it was done the first time, and taking good notes will make this possible. These notes are also needed to write laboratory reports explaining your results. For all of these reasons, your notebook needs to be thorough and accurate. As a general rule, a good notebook is one from which someone else can repeat your experimental work in the same way that you have done it.

3.1 GENERAL GUIDELINES

1. The laboratory notebook must be composed of pages attached within a binding. The pages must be numbered, and if your instructor requires it, may have a pressure-activated copy page ("carbon copy" or carbonless copy).

2. Name, course, and lab section number must be written on the cover or front page of your lab notebook.

3. Make sure your handwriting is legible.

4. Always use permanent ink, not pencil.

5. Notes must be written at the time the observations are made. Write it down NOW!

6. Write all data in your notebook—weights, temperatures, everything. When recording experimental data, always include units.

7. Use complete sentences, or at least enough words so that someone else can follow the train of thought.

8. Do not erase. If you make an error, draw a single line through it, add initials and date, and continue. The original statement should still be legible.

9. If you are using a lab notebook with pressure-activated copy ("carbon copy"), follow the direction of your instructor on which pages (original or duplicate) can be removed when submitting notebook pages for grading. Never remove both the original and copy pages from your notebook. You should always keep a backup record of your work.

10. Date every page as you use it.

There is considerable variation among lab instructors in how they address the general guidelines listed above. Naturally, you will need to incorporate specific instructions as given by the lab instructor of your course. While the lab notebook information described here is geared toward the use of a hardcopy notebook, most of these general guidelines also apply if you are using an electronic laboratory notebook. The key is to have a record that is informative, contemporaneous, and permanent.

The first several pages of your notebook should be reserved for a Table of Contents. From there, each experiment recorded in your notebook should contain sections A–G, which are outlined below. Sections A–G contain information that should be entered and completed prior to the laboratory period in which you begin the experiment.

3.2A Title

Start each experimental write-up with an accurate, descriptive title.

3.2B Purpose

Discuss the general purpose of the experiment in at most two or three sentences. This should include a statement of the hypothesis to be tested, or a question that will be answered if the experiment is successful. If the experiment is a synthesis (as opposed to a technique), write the chemical equation, including reagents and expected product(s). For multistep syntheses, write one equation for each transformation, including the preparation of reagents.

3.2C References

Cite the reference(s) you use in preparing your notebook. This includes the sources of physical properties and hazards of any compounds you'll use, as well as the source upon which your experimental procedure is based. In most cases this will be the lab textbook and/or your lab instructor's in-house course materials. While the sources of information may seem obvious while you're in this course, it's good to build a regular practice of including the references in your notebook.

3.2D Chemical Properties

Make a table that lists the chemical properties of all reactants and reagents that you will be using in the experiment. This table should include the name of each compound, its molecular weight (MW), density (d), melting point (mp), and boiling point (bp). Boiling point should include the pressure at which it was measured, if it wasn't under standard atmospheric pressure. Add the source or sources of these data to the References section of your notebook. Along with the properties, you should add a column where you list the amounts of each component you will use. The amounts should be expressed in terms of the units you measure (g or mL). For the reactants, include the moles, so that you can identify the **limiting reagent**. The molar equivalents of any reagents should also be listed; this will help you predict what unreacted materials may remain after the reaction is complete, so that you can handle and dispose of them properly. A **molar equivalent** is simply the ratio of moles of reagent divided by moles of the limiting reactant.

<< **limiting reagent**
The reaction component that is consumed first when other reactants are present in larger amounts. Many organic reactions have 1:1 stoichiometry, where 1 mol of reactant theoretically gives 1 mol of product; in these cases the reactant of lowest molar quantity is the limiting reagent.

<< **molar equivalent**
The ratio of moles of a reactant or reagent relative to the moles of limiting reagent, without adjusting for the stoichiometry of the reaction.

3.2E | Safety Guidelines

Make a table listing the safety hazards of the compounds that you will use, including solvents. For each compound, list the toxicity (if known), the flash point (in °C), and any other important safety information (e.g., flammable, corrosive, irritant, etc.).

The chemical properties (section 3.2D) and safety guidelines (section 3.2E) may be combined into one table.

3.2F | Equipment

Sketch any equipment setups or apparatus that you will use for the first time. Include in your drawing the positions of any clamps that are used. If you have already drawn the apparatus for an earlier experiment, you need only indicate the page in your notebook where the drawing can be found.

3.2G | Pre-Lab Flowchart

The pre-lab flowchart is a visual depiction of the order of steps you'll do in the lab, and the connections between the steps. It will help keep you organized and efficient in the lab, and will also help with identifying desired products versus waste materials. For more details, see section 3.4.

3.2H | Experimental (during Lab)

This section of your notebook is written during the course of a laboratory period. It is a record of what you do as you do it, and it must be completed before you leave the lab for the day. Some instructors may require portions of these records to be submitted at the end of the lab period, so be sure to keep the following records during your lab work:

1. As you move through the experiment, note any deviations from the procedure and flowchart. Such deviations might be intentional, in response to directions from the lab instructor, or an unintentional mistake that you will need to note in your records as an aid in interpreting the results.

2. Keep a log of both your actions and your observations. Any reader should be able to repeat the experiment as you ran it based on what you have written. Include any thoughts you have about what may be going on, or how the experiment might be changed in the future.

3. Make sure to record any melting points, boiling points, weights, etc., before you leave the lab whether you think you need them or not. Chances are that you will. Drawings of all thin-layer chromatography plates should also be included here. Alternatively, a hard copy print from a photograph may be permanently attached to the notebook page in place of a drawing.

4. Record your progress and observations completely and accurately. The information included here may help you understand later if your experiment was successful or what went wrong.

5. At the end of each day initial and date what you have written.

6. Follow the directions from your lab instructor about whether to submit these notebook pages before leaving the lab.

3.3 RESOURCES FOR CHEMICAL PROPERTIES AND SAFETY GUIDELINES

The following references (and many others) are available in either the laboratory or the library. Some are also available online. You should familiarize yourself with them, because you will use them frequently throughout the semester.

3.3A | Chemical Properties (See Section 3.2D)

1. *Aldrich Chemistry: Handbook of Fine Chemicals*; Sigma-Aldrich, 2007–2008 (or later edition). An online product search on the MilliporeSigma website (https://www.sigmaaldrich.com/US/en) provides access to the same information previously available in the handbook.

2. *CRC Handbook of Chemistry and Physics*, 99th ed. (internet version 2018); Rumble, J. R., Ed.; CRC Press/Taylor & Francis. Online at https://hbcp .chemnetbase.com.

3. *The Merck Index*, 12th ed.; Budavari, S., Ed.; Merck: Whitehouse Station, NJ, 1996 (or later edition). Online at https://www.rsc.org/merck-index.

3.3B | Safety Guidelines (See Section 3.2E)

1. *Safety in Academic Chemistry Laboratories: Best Practices for First- and Second-Year University Students*, 8th ed.; Finster, D. C., Ed.; American Chemical Society: Washington, DC, 2017.

2. Lewis, R. J. *Sax's Dangerous Properties of Industrial Materials*, 11th ed.; Wiley: Hoboken, NJ, 2004. Online at https://onlinelibrary.wiley.com/.

3. Safety Data Sheets are provided by chemical suppliers; for example, MilliporeSigma (https://www.sigmaaldrich.com/US/en).

3.4 PRE-LAB PREPARATION: THE PRE-LAB FLOWCHART

Working in the organic chemistry lab requires some advance preparation to formulate a "game plan" to organize your activities in the lab. This makes the activities safer and more efficient.

3.4A | What Goes in the Pre-Lab Flowchart?

An organized plan is important for each experiment that involves hands-on work with chemicals. While the pre-lab flowchart described here is one way to organize your experimental plan, you should check with your lab instructor to see if they have an alternative way to organize the experimental plan.

The flowchart is a visual representation of the inputs and outputs from the various operational procedures during an experiment, and it provides an abbreviated road map to help you know what to do next in the lab. Preparing the flowchart will help you to think more deeply about the experiment you are doing—not only about the reaction itself, but also the practical reasons behind each of the procedures you perform. The flowchart, then, will help to maximize what you can learn from each experiment. If you are unsure about the reasons behind a specific step or action that is required in an experiment, ask questions!

3.4B | Some General Instructions

1. Write a balanced equation for the reaction you will perform, if appropriate. A balanced equation will include any by-products of the reaction that will need to be separated. It is much easier to figure out what each step accomplishes if you know all of the reactants and products.

2. Provide a couple of words or a brief phrase to denote each operation and input/output, and don't go overboard with excessive detail. In most cases, a single page will be sufficient for the entire flowchart.

3. Write the flowchart in your notebook and submit a copy of it, using the submission method your instructor requires.

3.4C | Example of a Flowchart Created from an Experimental Procedure

A representative experimental procedure, "Synthesis of (*R*)-(+)-3-Methyladipic Acid," is described below in text format.[1] The reaction is shown in **Figure 3.1**, and a sample flowchart is provided (**Figure 3.2**), generated from the text instructions. As you read the text, examine the flowchart to see how it illustrates the same information in a graphical format. There is a pathway showing how to move through the various steps of the experimental procedure, and branch points that help track different materials that are generated. Every experiment is different, though, so the number of steps and branch points will vary. The flowchart you submit may not look exactly like the one in the sample.

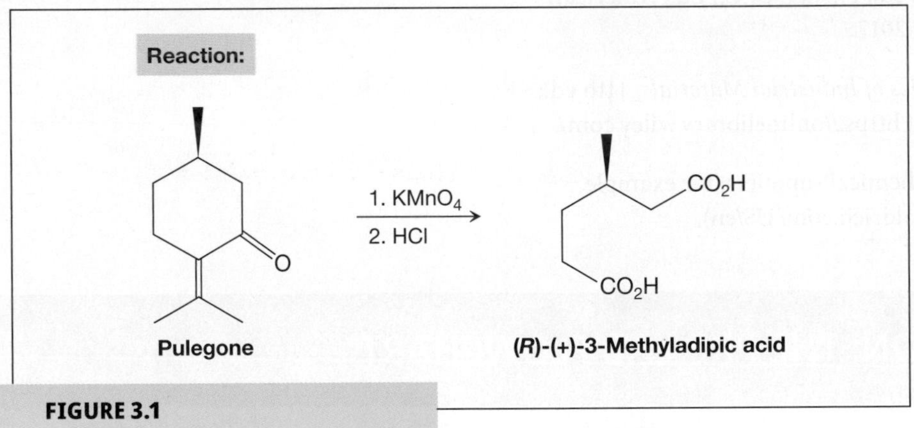

It's worth noting that this example synthesis is an older procedure that uses a lot of $KMnO_4$. Pay attention to the types and amounts of wastes that are generated, and how they are handled. The flowchart will help you determine whether or not this procedure does a good job of satisfying green chemistry principles (see Chapter 1). As you proceed through your own work in the lab, you can use a flowchart for a similar review of how well your own experiments address green chemistry principles.

Figure 3.3 shows a balanced equation of the reaction steps.

To a 250-mL Erlenmeyer flask containing 40 mL of distilled water, add 5 mL of pulegone. Swirl to mix the components (a two-phase mixture should result), then add 5 g of $KMnO_4$. Continue swirling for 10 minutes, and then allow the mixture to stand for approximately 2 hours, swirling occasionally. After this time, heat the mixture in a boiling water bath for 10 minutes. *CAUTION: The oxidation of pulegone is*

Reaction:

Pulegone

1. $KMnO_4$
2. HCl

(*R*)-(+)-3-Methyladipic acid

FIGURE 3.1

---

[1]Scott, W. J.; Hammond, G. B.; Becicka, B. T.; Wiemer, D. F. Oxidation of (*R*)-(+)-Pulegone to (*R*)-(+)-3-Methyladipic Acid. *J. Chem. Educ.* **1993**, *70*, 951–952.

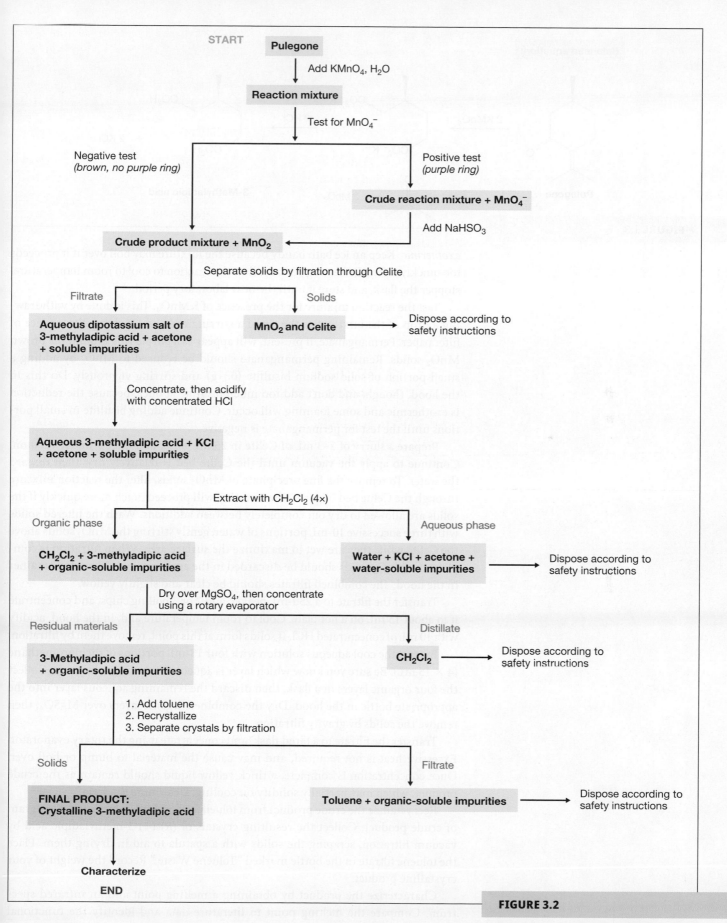

FIGURE 3.2

Representative pre-lab flowchart: synthesis of (R)-(+)-3-methyladipic acid.

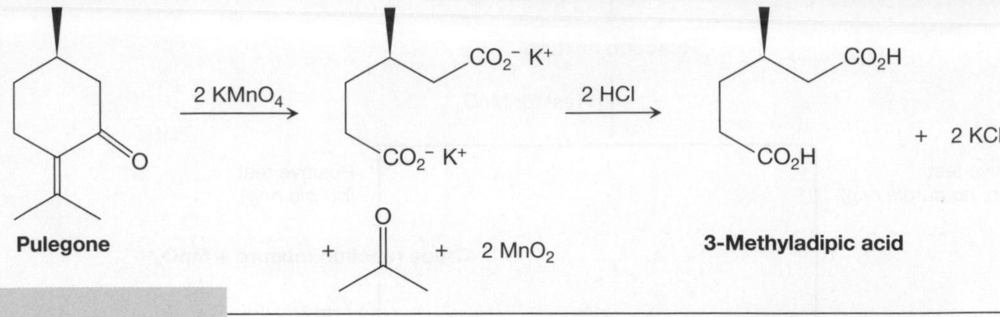

FIGURE 3.3

exothermic. Keep an ice bath handy because the mixture may boil over if it proceeds too quickly. After the heating period, allow the reaction to cool to room temperature, stopper the flask, and store it until the next laboratory period.

Test the reaction mixture for the presence of $KMnO_4$. This is done by withdrawing a drop of the mixture on the tip of a stirring rod and touching it to a piece of filter paper. Permanganate, if present, will appear as a purple ring around the brown MnO_2 solids. Remaining permanganate should be reduced to MnO_2 by adding a small portion of solid sodium bisulfite (0.1 g) and stirring vigorously. Do this in the hood, though, and don't add too much bisulfite at once, because the reduction is exothermic and some foaming will occur. Continue adding bisulfite in small portions until the test for permanganate is negative.

Prepare a slurry of 2–3 mL of Celite in 25 mL water and filter it under vacuum. Continue to apply the vacuum until the Celite bed is relatively dry, then discard the water. To remove the fine precipitate of MnO_2 solids, filter the reaction mixture through the Celite bed in parts. The filtration will proceed much more quickly if the solids are allowed to dry out completely between additions. Wash the filtered solids with three successive 10-mL portions of water, gently stirring the MnO_2 solids above the Celite while they are wet to maximize the surface area. When filtration is complete, the MnO_2 solids should be discarded in the appropriate solid waste container in the hood. The combined filtrates should be clear and slightly yellow.

Transfer the filtrate to a 250-mL beaker, add 2 or 3 boiling chips, and concentrate it to about 15 mL on a hot plate. Cool to room temperature and, in the hood, acidify with 10 mL of concentrated HCl. If solids form at this point, remove them by filtration.

Extract the cool aqueous solution with four 15-mL portions of dichloromethane (4 × 15 mL). Be sure you know which layer is aqueous and which is organic. Collect the four organic layers in a flask, then discard the remaining aqueous layer into the appropriate bottle in the hood. Dry the combined organic layers over $MgSO_4$, then remove the solids by gravity filtration.

Transfer the filtrate to a tared flask, and concentrate using the rotary evaporator. Excessive heat is not required, and may cause the material to bump or boil over. Once concentration is complete, a thick yellow liquid should remain as the crude product, which may partially solidify on cooling. Determine the yield.

Recrystallize the crude product from toluene, using 5 mL toluene for every gram of crude product. Collect the resulting crystals of (R)-(+)-3-methyladipic acid by vacuum filtration, scraping the solids with a spatula to aid in drying them. Place the toluene filtrate in the bottle marked "Toluene Waste." Record the weight of your crystalline product.

Characterize the product by obtaining a melting point and an infrared spectrum. Compare the melting point to literature data and identify the functional groups in your infrared spectrum.

Calculate your percent yield based on the amount of pulegone you used.

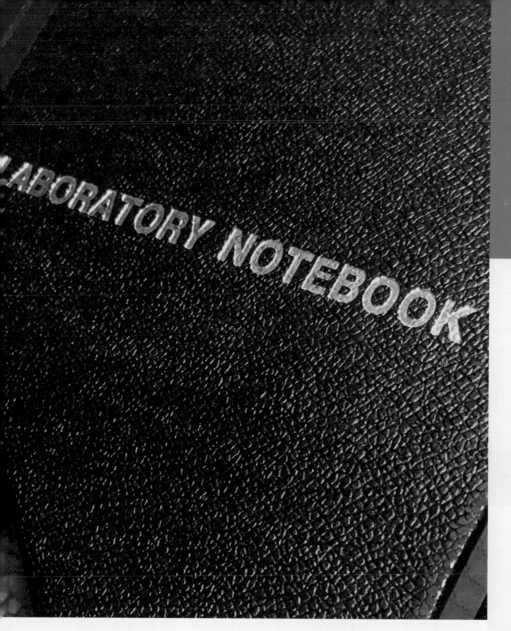

LEARNING OBJECTIVES

- Demonstrate your understanding of the importance of technical writing for various audiences.

- Report results and discussion in a format that clearly communicates them to your audience.

- Practice technical writing, using the components typically found in chemistry research journal articles.

The Laboratory Report

LABORATORY NOTEBOOK

Keeping an informative, contemporaneous, and permanent notebook that records your laboratory observations and data provides the foundation for an excellent laboratory report.

Courtesy of Gregory K. Friestad.

INTRODUCTION

After you have completed an experiment, including data analysis and interpretation, you will need to share your work with others. An essential component of doing science is communicating your experimental results to those outside the lab via written reports. Different audiences may require different writing styles. The audience could be scientists with a background in some other discipline, or members of the general public who happen to be interested in the field. Or, they could be experts in the same type of science, working on similar projects in a competing lab somewhere else in the world. Your audience could be your boss, too, who is about to make a case to the board of directors that your salary is due for a raise. Regardless of the audience, you need to write clearly and effectively about the work you have done in the lab.

If only an informal report is needed, then simply copying notebook pages and attaching them to a brief summary of the findings may suffice. In many cases, however, a more professional, technical report is needed, so a formal lab report must be submitted. Formal scientific writing takes experience and practice, and fortunately for you, you will have a number of opportunities to write formal lab reports for this course.

4.1 FORMAT

Technical writing is usually expected to follow certain formatting requirements that are imposed by journal publishers or by convention within the workplace. Similarly, when your instructor assigns a formal lab report, there is a specific format that is *required*. General formatting guidelines provided in this chapter should be followed closely unless your lab instructor gives separate directions. Your instructor may include additional requirements regarding font size, line spacing, margins, pagination, etc.; follow your instructor's requirements.

A typical report will be three or four pages long. These reports should be typed using the format outlined below. It should be complete, but concise. In the chemical literature, journals often have page limits for reports, although appendices such as data tables, spectra, copies of lab notebook pages, and other forms of data are not included in the page limit. Your lab instructor will notify you of any page limits for your reports.

Some rules for writing are commonly accepted for scientific literature and should be followed when generating your report. They are as follows:

1. Write in the past tense when describing what you did.

2. Some instructors prefer that you write in an active voice: "Recrystallization gave the pure product." Others prefer that you write in passive voice: "The product was purified by recrystallization." Use your lab instructor's advice on this point. However, you should *not* write in the first person: "I purified the product by recrystallization."

3. Incorporate data into the text of the report when explaining things, even if the data are already in a table. This helps to clarify the discussion.

4. Write with the three C's in mind: clear, concise, and complete.

Lab reports (and research journal articles) contain specific sections. These are explained here.

4.2A | Title Page

a. Title

b. Experiment number

c. Identifying information such as your name, section number, TA, and report submission date.

4.2B | Purpose

Discuss the general purpose of the experiment in at most two or three sentences (e.g., "The purpose of this lab was to investigate and compare various methods of distillation."). This should be more than a simple restatement of the title. If you are performing a synthesis, include the **balanced chemical equation**. The chemical equation provided in a textbook is not always balanced. Organic chemists often omit inorganic or other by-products such as HCl or CH_3OH from an organic reaction equation involving more complicated organic structures. In the lab, though, for safety reasons you must plan for how to contain or handle these by-products, so the balanced equation is essential.

<< **balanced chemical equation**
An accounting of all the atoms of reactants, defining how the atoms are distributed in products. The mole ratios of components in such an equation is called the stoichiometry.

4.2C | Experimental

During the course of the experiment, you should have recorded in your notebook a clear written account of the procedures that you followed. That will become your experimental section. It will be a concise description of what you did and how you obtained the data you present, sticking to the facts only (interpretations and conclusions come later). In some cases, if you have not deviated from a procedure from the textbook, your instructor may suggest that you cite that as a source rather than merely copying the procedure. You should indicate where the source information can be found in your notebook pages. Attach these pages to your report as an appendix.

4.2D | Results and Discussion

This is the most important section of the report. Here you should interpret your own experimental data and discuss what those data mean. You will need to explain clearly how you interpreted the results. It usually works best to divide this section into parts that correspond to the individual components of the experiment (e.g., each synthetic step). Do not include a detailed experimental account—this should already be written in your notebook pages. Instead, concentrate on your results. In a sentence or two, tell what you did and the outcome (e.g., "Vacuum filtration provided the product, benzoic acid."), then give the supporting evidence of that outcome (e.g., "The infrared spectrum showed a strong peak at 1689 cm^{-1}, indicating that the product may contain a functional group with a conjugated $C{=}O$."). Give all the evidence you have that a particular step was successful (or was not!). Not all experiments will use the same types of supporting evidence. If you have done a synthesis type of experiment, the following are some examples of supporting evidence that should be included in your report:

- Amount (in grams) and percent yield.
- Physical appearance: color, state, etc.
- Physical properties: melting point or boiling point, with literature value (cite your source) for comparison. Melting points should be reported as a range, as this provides evidence of purity.
- Thin-layer (TLC) or gas (GC) chromatographic data. For TLC, report the R_f value(s) and the solvent system used (drawings of TLC plates should be in your notebook). For GC, report the retention times and relative amounts of each component, and comment on the number of other peaks present, if relevant (the GC output, clearly labeled with experimental conditions such as column type, flow rate, and temperature, should be attached in an appendix).
- Spectroscopic data: infrared (IR) and/or nuclear magnetic resonance (NMR) spectra. List these peaks in the order they appear from left to right in the spectrum. For IR data, report key diagnostic absorbances only (in cm^{-1}), along with an assignment of the functional group of each absorbance. For NMR data, report the position in ppm, the integration (number of protons), and the multiplicity (e.g., 3.6 ppm, ^3H, triplet), and assign each signal to a proton (or protons) in the compound. Coupling constants for multiplets, if required, would also be included here (e.g., $J = 6.7$ Hz). Include a drawing of your compound (number the carbon atoms) in this section to facilitate identification and interpretation of specific resonances. If more than two or three peaks are reported, present them in a table. Examples of such tables are given in the corresponding spectroscopy chapters, Chapters 6–8. Explain the significance of the observed peaks, and note expected peaks that are absent, too, where relevant. For example, if an alcohol is being oxidized to a ketone, and no C=O peak is observed in the IR spectrum, this should be reported because it is important evidence that the reaction did not work as expected. In an appendix, attach a hard copy of each original spectrum, clearly labeled so that a reader knows at which point each spectrum was obtained (i.e., before recrystallization or after).

Once you have presented the results and supporting data, draw specific conclusions based on the data that *you* obtained. For a synthesis type of experiment, these will focus on identity and purity of the product: Did you really make the compound or not? Did the distillation work or not? Is your product pure? If not, does the evidence suggest what might have happened? The answers to these kinds of questions *must* be consistent with *your* data, and *must* be supported with experimental evidence *from your own work* as compared to literature-based expectations. Clearly explain how you drew your conclusions. Such evidence should include a comparison of the physical and spectral properties of your product with those of the starting material, and with literature values (with citation) expected for the product whenever possible. Discuss the yield of the product you obtained. If it is unreasonably low, then suggest possible explanations.

4.2E Conclusion

Summarize the main points in the Results and Discussion section, and indicate whether the Purpose was accomplished. Was your experiment successful? Where appropriate, discuss what your specific outcome tells you about a general theory or class of reactions. Did any of your results agree or disagree with your expectations? If a new technique was utilized, comment on its effectiveness relative to what you were trying to accomplish.

4.2F References

Provide citations to all of the references you used in preparing, carrying out, and analyzing the experiment. This includes the sources of procedures, the references you consulted in analyzing the data, and primary or secondary literature you used for comparisons of melting and boiling points, IR spectra, NMR spectra, and the like. Chapter 9, on the chemical literature, provides extensive examples of how to properly cite your sources.

4.2G Appendices

You may not always have material for each of the appendices listed here, but the material you do have must be labeled and appear in the following order:

1. Appendix A: Calculations (percent yields, R_f values, etc.). For multistep synthesis experiments, include a percent yield calculation for each synthetic step. In other cases, where multiple calculations of the same type are done, a sample calculation is sufficient. Include the equation you utilized, complete with units.

2. Appendix B: Spectra (IR, NMR), GC printouts, etc.

3. Appendix C: Experimental records (notebook pages), including any pictures or drawings you have of your experimental setup.

If your lab instructor has assigned any post-lab questions, be sure to attach the questions and your answers to them.

4.3 EVALUATION OF LAB REPORTS

The quality of a lab report is based on (a) following the required format, (b) including the relevant data, and (c) interpreting the appropriate data to draw conclusions of relevance to the purpose of the experiment. All of these are independent of whether the experiment worked well.

Still, this is a laboratory course. So, experimental results are typically a component of the grade. It is likely, then, that your instructor may consider your experimental success, such as the quality and amount of the product, in the grading evaluation. Some instructors may require a sample of the product to be submitted, in addition to the report; follow your instructor's instructions on this.

Regardless of how successful or unsuccessful the experiment seems to have been, a well-written report will still show significant gains in knowledge of practical organic chemistry. This is true even in cases where poor yield or purity was observed. In such cases, you should still put in the effort to write a strong report, because even unsuccessful experiments can be useful tools to meet the learning objectives. Most instructors place more value in achieving the learning objectives rather than focusing solely on the yield.

5

LEARNING OBJECTIVES

- Explain how separation techniques purify compounds.
  - » *Explain how filtrations separate solids from liquids.*
  - » *Explain how extractions separate compounds of differing solubilities.*
  - » *Explain how distillations separate liquids with different boiling points.*
  - » *Explain how recrystallizations purify organic solids.*
  - » *Explain how chromatography separates compounds for analytical and preparative purposes.*
- Select the appropriate purification methods for different types of mixtures.
- Explain how boiling points, melting points, densities, and optical rotations are measured.
- Use measurements of physical properties as evidence of identity and purity.

Purifications of Organic Compounds and Determination of Their Physical Properties

PANNING FOR GOLD

Separation techniques for mixtures in the organic lab depend on the physical properties of the components. Here, a specialized pan aids in using water to separate heavier gold particles from lighter particles of dirt or gravel.

Jeffrey B. Banke/Shutterstock.

INTRODUCTION

The organic chemistry laboratory is where new materials are made—materials with biological or physical properties that are both interesting and practical. Preparing these materials, purifying them, providing evidence of their identity, and measuring their properties are all activities you will experience in this course. There are a variety of techniques to synthesize, purify, characterize, and measure the compounds, which depend on a compound's characteristics. This chapter introduces you to some of the fundamental tools needed in the lab, and should be consulted regularly as a general resource for your work throughout the term.

5.1 SEPARATION OF LIQUIDS AND SOLIDS BY FILTRATION

Filtration is the physical separation of solid and liquid phases—familiar to anyone who has made a cup of coffee. It can be carried out in a number of ways, depending on the nature of the compounds. Generally, the mixture of solids and liquid is poured into a funnel, where it passes through a porous material (cloth, filter paper, or porous glass disc) that retains the solid phase, allowing the liquid phase to pass through. Different techniques and glassware are used, depending on whether the flow of liquid is aided by gravity or vacuum.

5.1A Gravity Filtration

Gravity filtration tends to be used when the liquid phase contains the material of interest, and the solid is an undesired material. Examples include removing the **drying agent** $MgSO_4$ from an organic solution, removing activated charcoal from a decolorized solution, or removing an insoluble side product from a reaction mixture.

The key to carrying out this process efficiently is to use a fluted filter paper in a simple conical funnel with an open stem (**Figure 5.1a**). A filter paper is "fluted" by folding in half, then folding in small wedge-shaped segments in alternating directions with each fold passing through a point at the center of the original circle shape (**Figure 5.1b**). Upon opening, all the creases radiate outward from the center of the filter paper, providing increased surface area for the free flow of solvent through all parts of the paper when it is placed into a conical funnel (**Figure 5.1c**). If the filter paper is made into a cone shape without the fluting, it will lie flat against the inner walls of the funnel and filtration will occur mostly at the tip of the filter paper; this is inefficient, particularly when filtering fine particulates, which can clog the filter paper.

Occasionally the solid is so finely divided that it even clogs the pores of a properly fluted filter paper, causing the flow to be unacceptably slow. In such cases, **diatomaceous earth** ("Celite" or "Filter-Aid") may be employed; it helps to distribute fine particles so that they don't accumulate in the filter paper. It is usually wetted to form a slurry, using the same solvent present in the mixture, and then filtered first. This leaves a firm pad of Celite on the filter paper prior to adding the mixture. Some Celite may also be added directly to the mixture to be filtered; as the resulting slurry is poured onto the pad of Celite in the funnel,

drying agent >>
A salt that rapidly adsorbs water to form a hydrate. In its dehydrated form, a drying agent is added to organic solutions in order to remove water.

diatomaceous earth >>
Sedimentary deposits from the fossilized remains of diatoms. In powdered form, this material is insoluble in organic solvents and is used to aid in filtration, preventing fine particulates from clogging the filter. Also known as Celite or Filter-Aid.

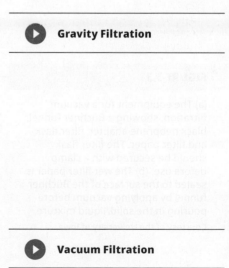

FIGURE 5.1

(a) Apparatus for gravity filtration. (b) Preparing a filter paper for gravity filtration. (c) Hot filtration. Secure clamps or tongs may also be used for safely handling hot glassware.

(a)

(c)

Erlenmeyer flask

(b)

particulates will be dispersed throughout the Celite, helping to avoid clogging the filtration.

Recrystallization (section 5.4) is a purification process in which the desired product forms crystals from solution. Gravity filtration is often used during recrystallization, before the crystals form, to remove undesired insoluble materials. This procedure is often called "hot filtration" (Figure 5.1c). In this case, the conical funnel is placed into an Erlenmeyer flask containing a small amount of solvent, and the apparatus is placed on a hotplate and warmed by heating to a gentle boil. An initial rinse of the filter using a small amount of the hot solvent mixture can help bring the apparatus up to temperature. Keeping the entire apparatus warm as the filtration proceeds allows the desired compound to stay in solution so that it does not crystallize in the filter paper along with the unwanted solids.

▶ **Gravity Filtration**

5.1B   Vacuum Filtration (Suction Filtration)

Vacuum filtration should be used when the material of interest is the solid, not the liquid. It is used during a recrystallization to recover the desired compound after it has cooled and crystallized. A filter paper is chosen with a proper size to lie flat upon the porous surface inside a Büchner or Hirsch funnel (**Figure 5.2**), trimming excess paper if necessary. For small Hirsch funnels, this means a 1-cm filter paper (about the size of a thumbnail).

▶ **Vacuum Filtration**

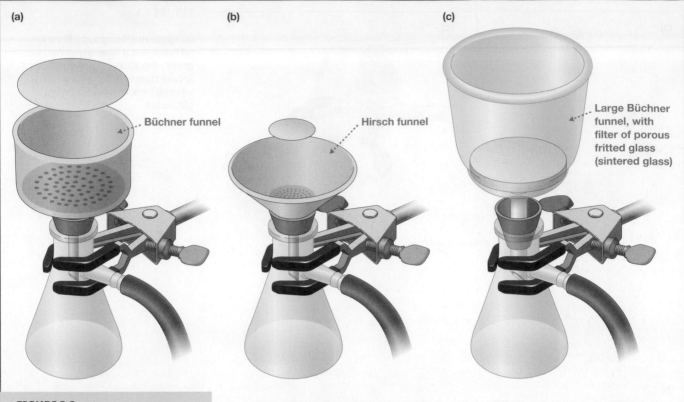

(a) (b) (c)

Büchner funnel

Hirsch funnel

Large Büchner funnel, with filter of porous fritted glass (sintered glass)

FIGURE 5.2

(a–c) Büchner and Hirsch funnels used in vacuum filtration. Smaller Hirsch funnels, which hold a 1-cm filter paper, are preferred for small amounts of material.

During filtration, the funnel rests in the top of a filter flask (an Erlenmeyer flask with a vacuum sidearm), with a conical neoprene adapter in between, which provides a seal when vacuum is applied by connecting to the vacuum source with a thick-walled hose (**Figure 5.3**). The filter paper is wetted first with the same solvent as the mixture you are filtering, and vacuum is applied to hold down the filter paper before the mixture is added to the funnel; this prevents solids from bypassing the filter paper. After the filtration is complete, the solid can be washed with an additional portion of cold solvent, and then dried by using the vacuum to pull air through the solid.

FIGURE 5.3

(a) The equipment for a vacuum filtration, showing a Büchner funnel, black neoprene adapter, filter flask, and filter paper. The filter flask should be secured with a clamp before use. (b) The wet filter paper is sealed to the surface of the Büchner funnel by applying vacuum before pouring in the solid/liquid mixture.

Courtesy of the University of Iowa.

(a) (b)

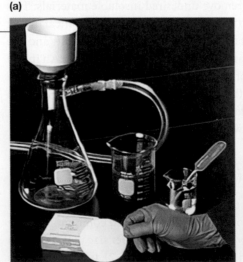

5.2A Background

Liquid–liquid extraction is designed to partition compounds between two liquid phases. Organic reactions are often followed by a **workup** that includes an extraction to separate water-soluble by-products from organic products. Extraction is particularly useful in the separation of water-soluble acidic and/or basic components from an organic mixture.

The success of liquid–liquid extraction depends on two factors. First, the two solvents must be *immiscible*, or not mutually soluble, so that they form two separate layers. Second, solutes must have different solubilities in the two liquid phases. Thus, an organic compound in the presence of two immiscible solvents, such as water and diethyl ether, will distribute (partition) itself between the two phases until equilibrium is reached. Typically, organic compounds are mostly in the organic (diethyl ether) phase, and ionic compounds (salts) are in the aqueous (water) phase.

At equilibrium, the ratio of concentrations of the solute in each layer is constant, and may be defined as the **partition coefficient** or the *distribution coefficient*, K:

$$\text{Distribution coefficient } (K) = \frac{[X]_B}{[X]_A} = \frac{\text{solubility of X in solvent B}}{\text{solubility of X in solvent A}}$$

where $[X]_B$ is the concentration of solute in solvent B (generally the organic phase) and $[X]_A$ is the concentration of solute in solvent A (generally the aqueous phase). This relationship is independent of the total concentration of the solute and the actual volumes of the two solvents. The distribution coefficient has a constant value for each combination of solute and solvent. In pharmaceutical chemistry, this coefficient is frequently cited as the "logP" of a drug, and is measured using octanol and water as the two-phase system. The logP reflects the **lipophilicity** of a drug, which has an impact on how it is absorbed in the body, where it is concentrated, and how it is eliminated.

Imagine that you have a solution of some solute (S) in solvent A, and it is mixed with a second solvent B, which is immiscible with A. The solute can be transferred between phases in an equilibrium process (**Figure 5.4**). Eventually the solute will reach equilibrium at K. The layers are separated. If the distribution coefficient K is very large ($K > 100$), the solute is virtually all found in solvent B. More commonly, though, not all the solute will be transferred in a single extraction. However, one or two additional extractions with a small amount of solvent B will recover whatever solute may have remained in solvent A after the first extraction. Extracting with two or three smaller portions of solvent is more efficient than a single extraction with a larger amount of solvent. Generally speaking, any organic compound with $K > 1$ can be efficiently extracted from aqueous solution.

<< **workup**
A procedure for separation of an organic product from other materials after a reaction is complete, often involving deactivation or quenching of excess reagents, liquid–liquid extraction, and solvent removal. A two-phase aqueous–organic extraction in a separatory funnel is commonly used to separate organic products from water-soluble by-products.

<< **partition coefficient**
A ratio that describes the portion of a compound dissolved in an organic phase versus an aqueous phase when the compound is mixed with both phases and allowed to reach a solubility equilibrium. Also known as the distribution coefficient.

<< **lipophilicity**
A term describing the degree of solubility in an oil-like solvent.

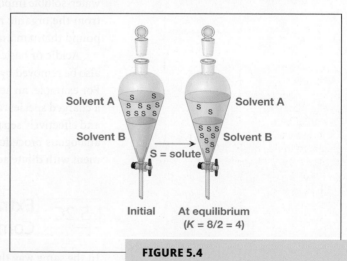

FIGURE 5.4

Consider a 10-g sample of a compound with $K = 4.0$, partitioned between organic and aqueous phases. At equilibrium distribution, four parts ($10 \text{ g} \times 4/5 = 8.0 \text{ g}$) of the compound will be in the organic phase and one part ($10 \times 1/5 = 2.0 \text{ g}$) of it will be in the aqueous phase. With a single extraction of 60 mL of organic solvent, 8.0 g of the compound will be obtained by evaporating the organic phase, only an 80% recovery. However, by dividing the same 60 mL of solvent into three 20-mL portions, the total amount of compound recovered may be increased to > 99%:

$$\text{First extraction: } 10 \text{ g} \times 4/5 = 8.0 \text{ g}$$
$$\text{Second extraction: } (10 \text{ g} - 8.0 \text{ g}) \times 4/5 = 1.6 \text{ g}$$
$$\text{Third extraction: } (10 \text{ g} - 8.0 \text{ g} - 1.6 \text{ g}) \times 4/5 = 0.32 \text{ g}$$
$$\text{Total: } 9.92 \text{ g } (99.2\% \text{ recovery})$$

In cases where $K < 1$, a simple extraction process will not give a satisfactory recovery of organic solute from an aqueous solution. In this case, however, the distribution coefficient can be altered by adding an inorganic salt such as sodium chloride to the aqueous layer. Organic compounds are generally less soluble in a saturated salt solution than in water, so the addition of NaCl shifts the equilibrium of solute between the two phases toward the organic layer, thereby increasing the distribution coefficient, and increasing the efficiency of extraction. This process is called *salting out*. Conversely, water tends to move into the saturated salt (brine) layer to help solvate the inorganic ions. Thus, saturated aqueous salt solutions are frequently used as preliminary drying agents to extract water molecules from the organic layer into the aqueous layer.

5.2B Extraction to Obtain Organic Products from Reaction Mixtures

Extraction is an important tool for the preliminary purification of a reaction product, where inorganic by-products or organic salts can be washed away by extraction into an aqueous phase. In this way, the desired product of a reaction may be separated from unreacted starting materials, unwanted by-products, and impurities. In a typical extraction sequence, a two-phase system of water and an organic solvent (commonly diethyl ether or *tert*-butyl methyl ether) is utilized to remove unwanted water-soluble impurities (inorganic salts, low-molecular-weight polar organics, etc.) from the organic reaction medium. Usually the organic product is a neutral compound that remains in the ether phase.

Acidic or basic impurities that might otherwise remain in the organic layer can also be removed by formation of their corresponding water-soluble salts (**Figure 5.5**). For example, an acidic impurity (RCO_2H), in the presence of aqueous NaOH, forms a charged species, a *salt* ($RCO_2^- Na^+$), thereby making it soluble in the aqueous layer, and effectively separating it from the uncharged components in the organic layer. An analogous procedure can be applied for the removal of basic impurities upon treatment with dilute acid.

5.2C Extraction for the Separation of Organic Compounds

In the same way that acidic and basic impurities can be removed from a neutral organic compound, the acidic and basic properties of organic compounds can be utilized to separate the components of a mixture. Organic acids (carboxylic acids and

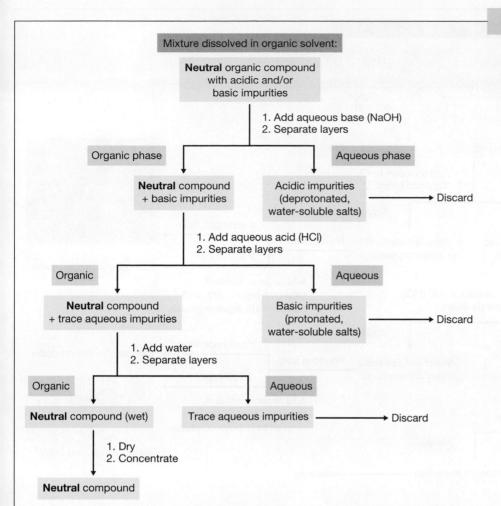

FIGURE 5.5

A general extraction sequence for the removal of acidic and basic impurities from a neutral organic compound. If one type of impurity (acid or base) is known to be absent, that portion of the sequence may be skipped.

phenols) and organic bases (amines) can be readily separated from each other and from neutral compounds by the extraction protocol outlined in **Figure 5.6**. Thus, stronger organic acids such as carboxylic acids ($pK_a \approx 5$) are easily converted into their sodium salts by reaction with sodium bicarbonate (for bicarbonate, H_2CO_3, $pK_a = 6.4$). Weaker organic acids such as phenols ($pK_a \approx 10$) require a stronger base such as sodium hydroxide. If both a phenol and a carboxylic acid are present, their differences in acidity allow their selective separation by extraction with the appropriate base. Aqueous sodium bicarbonate converts only the carboxylic acid to a salt, so only the carboxylic acid component is then drawn into the aqueous layer. If a phenol is present, it would be left behind in the organic phase because it is not acidic enough to be converted to a salt by $NaHCO_3$, and it could be later extracted into a more strongly basic NaOH solution.

Conversely, organic bases such as amines are converted into water-soluble hydrochloride salts by reaction with hydrochloric acid, and they may be separated from neutral and acidic substances by extraction with aqueous HCl.

Once the organic and aqueous phases have been separated, the various components can be isolated—by removing the solvent from the organic layer, and by neutralizing and subsequently filtering or extracting the aqueous layer. Using these principles it is possible to separate the various components from rather complex organic mixtures.

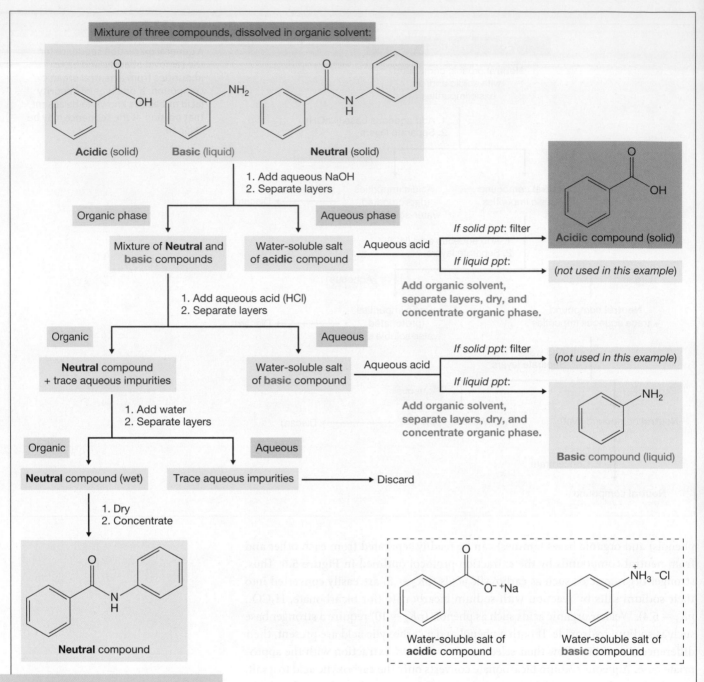

FIGURE 5.6

General approach to the separation of neutral, acidic, and basic compounds from a mixture. This is the same as Figure 5.5, except that the water-soluble salts are neutralized with acid or base so that the acidic and basic compounds can be recovered from the aqueous fractions. *Note:* ppt = precipitate; can be solid or liquid (oily droplets).

5.2D Practical Aspects of Extraction

Extraction is carried out in a separatory funnel (**Figure 5.7**), which is a cone-shaped enclosed funnel with a stopper at the wider end and a valve or stopcock at the narrower end. The separatory funnel rests in an iron ring with the narrow end down. The two liquid phases are placed in the funnel, which is closed, inverted with the narrower end up, swirled or shaken gently. Pressure may build up, especially with solvents having low boiling points, or if gas is produced by a reaction, such as that of NaHCO₃ (sodium bicarbonate) to make a salt from an acid, producing CO₂. Gas pressure is vented from the separatory funnel by opening the stopcock briefly while it is pointed upward, into a hood, and away from people. The separatory funnel is then placed back in the iron ring, and the stopper is removed from the top. The layers will separate, and the more dense layer can be drained off into an Erlenmeyer flask through the stopcock at the narrow end. As the phase boundary approaches the stopcock, begin to close the stopcock to slow the flow rate; this will avoid overshooting.

IDENTIFYING THE PHASES

An extremely important aspect of separations and extractions using a separatory funnel is the correct identification of the organic and aqueous layers. The easiest and most accurate way to predict which layer will be on top and which layer will be on bottom is to compare the densities of the two solvents being used. The denser layer will be the bottom layer, and the less dense layer will sit on top of it. Organic solvents such as hexane, diethyl ether, and ethyl acetate are less dense than water, while halogenated organic solvents such as chloroform and dichloromethane are more dense than water. However, the presence of other solutes in the aqueous phase, such as NaOH or NaCl, changes its density, so it is wise to confirm the identity of the organic and aqueous layers by adding a few drops of water to each; the water will dissolve only in the aqueous fraction.

If a layer is discarded before identifying it properly, an important component may be lost, and an experiment may have to be repeated from the start! Therefore, *save all the layers of any extraction until the product is isolated.* This way, if layers were incorrectly identified, they are all still available, and the product can be retrieved.

EMULSIONS

Sometimes a mixture that you know should separate into organic and aqueous phases fails to separate in the separatory funnel. There may appear to be a third, milky layer, or the entire mixture may appear to be one cloudy phase. This combined phase is called an **emulsion**, and it interferes with the extraction, making it hard to drain just one phase from the separatory funnel. To address this situation, add brine (saturated aqueous sodium chloride) to the mixture, mix it gently (shaking vigorously tends to make emulsions worse), and wait patiently. If partial separation occurs, drain out all the lower clear phase and repeat the separation with a fresh portion of the solvent in the lower phase.

Examples of liquid–liquid extractions. Worked Examples 1 and 2 represent common cases where compounds are separated based on their acid and base solubility properties.

Worked Example 1

A student has a mixture of *p*-toluic acid (pK_a = 4.36), *p-tert*-butylphenol (pK_a = 10.16), and acetanilide (pK_a = 22); see **Figure 5.8**. How can these components be separated from one another?

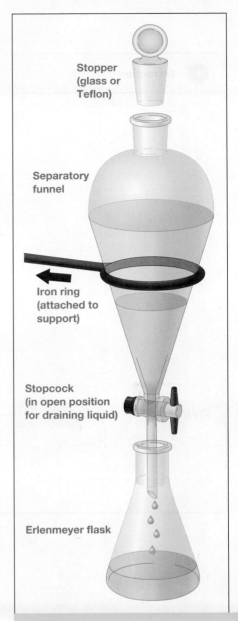

FIGURE 5.7

Separatory funnel.

<< emulsion
A nonhomogeneous mixture of two or more phases, such as organic and aqueous, that resists separation into two clearly defined layers.

FIGURE 5.8

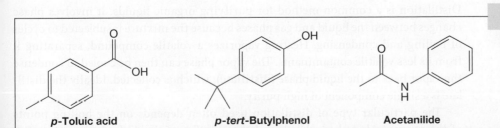

p-Toluic acid   *p-tert*-Butylphenol   Acetanilide

Solution

First, dissolve the mixture in organic solvent such as *tert*-butyl methyl ether or ethyl acetate. In this mixture, only *p*-toluic acid is acidic enough to be converted into a salt by HCO_3^-, so selectively extract it into aqueous $NaHCO_3$ solution. Next, extract the *p-tert*-butylphenol left behind in the organic phase using a more strongly basic aqueous NaOH solution. Separately neutralize each of the extracts by adding aqueous HCl, which will precipitate the water-insoluble *p*-toluic acid and *p-tert*-butylphenol. Then recover these two compounds by vacuum filtration. The third compound, acetanilide, is a neutral compound—not acidic enough to react with either NaOH or $NaHCO_3$ in aqueous solution—so it will remain dissolved in the nonpolar organic solvent throughout the sequence of extractions. Recover it by drying the organic layer with a drying agent such as anhydrous magnesium sulfate ($MgSO_4$) or sodium sulfate (Na_2SO_4), separating the drying agent by decanting or filtering (see Figure 5.1), and removing the organic solvent by concentrating it on the rotary evaporator. After all three of the components have been separated in this fashion, they can be recrystallized (section 5.4) to obtain pure substances.

Worked Example 2

A student is preparing an amide from 4-methylaniline and pentanoic acid via the acid chloride (**Figure 5.9**). Usually there are leftover carboxylic acid and amine that must be separated from the amide product. How should the student remove any leftover reactants from the desired amide product?

FIGURE 5.9

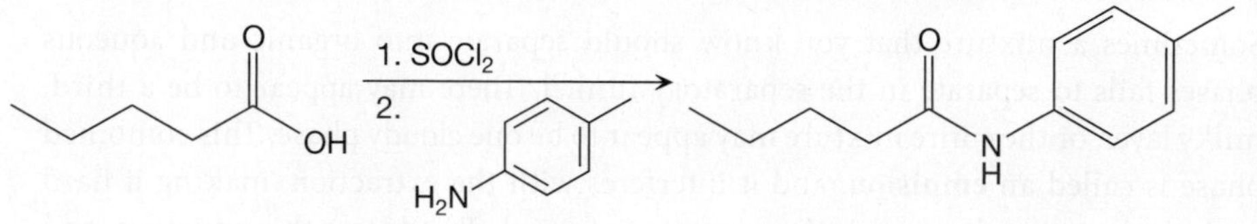

Solution

The amide product is neutral, whereas the leftover reactants are acidic (pentanoic acid) and basic (4-methylaniline). To remove any leftover reactants, dissolve the product mixture in an organic solvent such as ethyl acetate, then extract sequentially with aqueous HCl, aqueous NaOH, and water. Dry the organic phase over sodium sulfate, concentrate it on the rotary evaporator, and recrystallize (section 5.4) it to obtain the pure amide product.

5.3 PURIFICATION BY DISTILLATION

distillate >>
The liquid output collected from a condenser during distillation, a process of boiling and condensing that separates a liquid from other components of a mixture.

Distillation is a common method for purifying organic liquids. It involves phase changes between the liquid and gas phases because the mixture is subjected to cycles of boiling and condensing. Heating vaporizes a volatile compound, separating it from its less volatile contaminants. The vapor phase can then be cooled to condense the vapor back to the liquid phase (**distillate**), which is collected. Ideally the distillate is a single component of high purity.

The particular type of distillation used often depends on the boiling points of the compounds to be isolated, and the specifics of the desired separation. For example, *simple distillation* can easily separate two mutually soluble substances which differ in boiling points by 80°C or more. Simple distillation can sometimes be successful when boiling points are 40–80°C. When the differences in boiling point are less than 40°C, *fractional distillation* may be required.

5.3A | Simple Distillation

In a simple distillation apparatus (**Figure 5.10**), a distillation flask (or boiling flask) is connected to a distillation head, which is simply a vertical tube with a thermometer port and a downward-angled sidearm for distillate to drain out. Complete condensation from vapor back to liquid (the *distillate*) is ensured by attaching a jacketed condenser to the sidearm; compressed air or cold water is passed through the jacket to keep the condenser cool. A curved distillation adapter connects the condenser to a receiving flask, where the distillate collects, and also provides a vacuum or inert gas inlet.

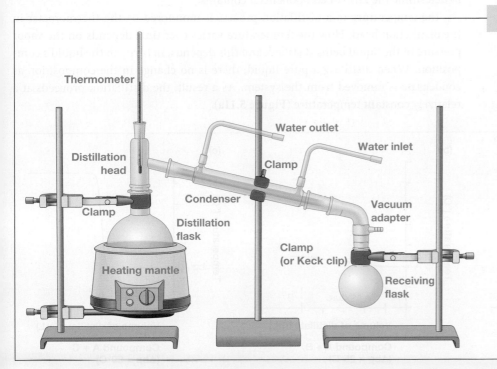

FIGURE 5.10

Simple distillation apparatus. The thermometer is inserted so that its reservoir bulb is below the sidearm of the distillation head. Heating may be supplied by sand bath, oil bath, or heating mantle (as shown). A magnetic stirrer may be added below the heat source.

It's important to clamp the apparatus securely. Two clamps should be used: The first at the neck of the distillation flask and the second on the condenser. Both clamps should be securely connected to a ring stand or built-in support bars in a fume hood. Be cautious in applying second and third clamps; tighten them only enough to support weight, and avoid twisting and pressure on other parts of the clamped apparatus, which can cause breakage. After the apparatus is set up and clamped, check carefully that all the ground glass connections remain tightly sealed. Sometimes it is necessary to switch the receiving flask during distillation; this is facilitated if a plastic clip ("Keck clip") is used in place of the clamp on the neck of the receiving flask.

The sample mixture is placed in the distillation flask by disconnecting the distillation head and pouring the liquid through the neck of the flask with the aid of a funnel. Alternatively, the liquid can be poured in through the opening created by removing the thermometer adapter; use a long-stem funnel to ensure that this mixture travels toward the distillation flask, not the receiving flask. Insert a boiling stick or boiling stone in the distillation flask, or use a magnetic stirrer, to facilitate smooth boiling with smaller bubbles. This prevents sudden eruptions of large volumes of vapors, known as "bumping," which can cause undistilled material to splash into the condenser, contaminating the distillate.

 Simple Distillation

A caution before heating: Never heat a sealed system—dangerous pressure can build up and cause an accident. Before you begin applying heat, make sure the vacuum adapter hose connector is unblocked so that the pressure can equalize with the external atmospheric pressure.

Heat is applied to the distillation flask, raising the temperature within, and increasing the vapor pressures of the components. When the sum of those vapor pressures reaches atmospheric pressure (or the pressure within the apparatus), the material in the distillation flask begins to boil. Vapors are carried upward to the distillation head, and their temperature is monitored. A cooled condenser slopes downward from the distillation head, ensuring that liquid condensing from cooled vapors will drain toward the receiving flask. As the distillation progresses, the receiving flask can be periodically changed, so that different fractions of distillate are collected in different flasks. Each fraction can be analyzed by gas chromatography to determine the ratio of components it contains.

The temperature of a distillation process is monitored by the thermometer in the distillation head. How the temperature varies over time depends on the vapor pressure of the liquid being distilled, and this depends, in turn, on the liquid's composition. When distilling a pure liquid, there is no change in the composition as condensate is removed from the system. As a result, the distillation proceeds at a relatively constant temperature (**Figure 5.11a**).

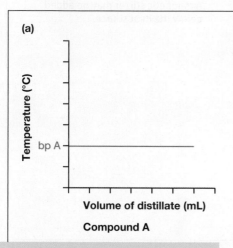

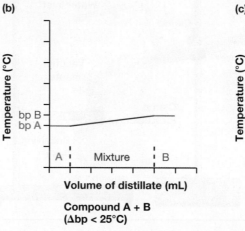

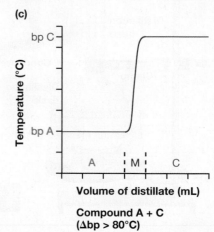

FIGURE 5.11

Temperature readings at the distillation head during a simple distillation of (a) a relatively pure compound, (b) a mixture of two components having boiling points (bp) differing by < 25°C, and (c) a mixture of two components with boiling points differing by > 80°C.

For any liquid mixture, though, the composition and temperatures change during the distillation. Dalton's law and Raoult's law describe this behavior.

Dalton's law: The vapor pressure of a liquid is the sum of the partial pressures of the individual components:

$$P = P_A + P_B$$

Raoult's law: At a given temperature and pressure, compound A has a partial pressure (P_A) in a mixture that is equal to the vapor pressure of the pure compound (P_A^{pure}) multiplied by its mole fraction (X_A) in the mixture:

$$P_A = (P_A^{pure})(X_A)$$

As the distillation proceeds, the composition of both the liquid and the vapor change; early in the distillation, the component of lower boiling point (greater partial pressure, P_A) is more rapidly removed from the system. As a result, the temperature increases throughout the distillation (**Figure 5.11b** and **Figure 5.11c**). A phase

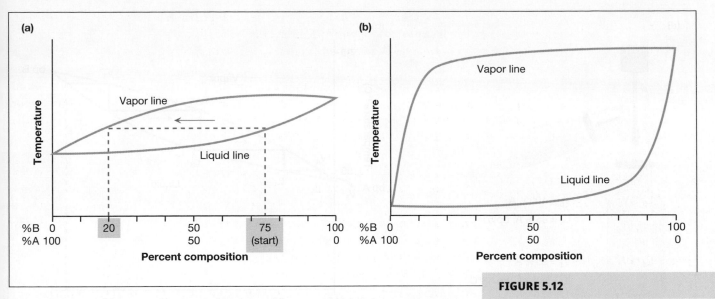

(a)

Temperature

Vapor line

Liquid line

%B 0   20   50   75   100
%A 100   50   (start)   0

Percent composition

(b)

Temperature

Vapor line

Liquid line

%B 0   50   100
%A 100   50   0

Percent composition

FIGURE 5.12

Phase diagrams for a mixture of two components (A and B) having (a) similar (< 80°C) boiling points and (b) widely differing (> 80°C) boiling points. In both cases, the boiling point of A is less than the boiling point of B.

diagram (**Figure 5.12**) is a plot of composition (mol %) versus temperature (T), where the lower curve is the liquid line and the upper curve is the vapor line. These diagrams help to explain the temperature-versus-volume behavior (Figure 5.11) of liquid mixtures, and additionally can be used to determine the composition of both the liquid and the vapor phase at any temperature throughout the distillation.

When examining a liquid–vapor phase diagram, keep in mind that the phase change from liquid to vapor is *not* accompanied by a change in temperature; the vapor and liquid are in equilibrium at the same temperature. However, the compositions of the vapor and liquid are different (Figure 5.12), which explains how distillation can change the ratio of the two components. The horizontal line represents the phase change at constant temperature from liquid to gas phase. This horizontal line intersects the liquid and vapor lines at two different compositions. The dotted lines in Figure 5.12 show that a boiling liquid starting at a ratio of 75% B and 25% A is in equilibrium with a vapor consisting of 20% B and 80% A. Condensing that vapor and cooling it gives a liquid which is enriched in A, the compound with the lower boiling point.

5.3B Fractional Distillation

In fractional distillation, an extra vertical column called a *fractionating column* is added to the apparatus (**Figure 5.13**). The physical principles that describe a simple distillation hold true for a fractional distillation as well, but the fractionating column offers a much larger surface area. The vapor undergoes a continuous cycle of condensation and revaporization as it passes up through the fractionating column, and each sequential revaporization is equivalent to another simple distillation. Thus, the composition of the vapor is progressively enriched as it moves up the column. As a result, the temperature behavior of a fractional distillation over time resembles that shown in Figure 5.11c, even when two components have similar boiling points. With an apparatus like the one shown in Figure 5.13, it is possible to cleanly separate components of liquid mixtures in which the boiling points differ by as little as 25°C. Compounds with boiling points closer than this can be separated, but it may require a more specialized apparatus, such as a longer fractionating column. It should be noted that the fractional distillation technique generally takes more time, and can result in greater material losses because of the extra glassware surface area, so it is used only when the boiling points are too close to separate by simple distillation.

(a)

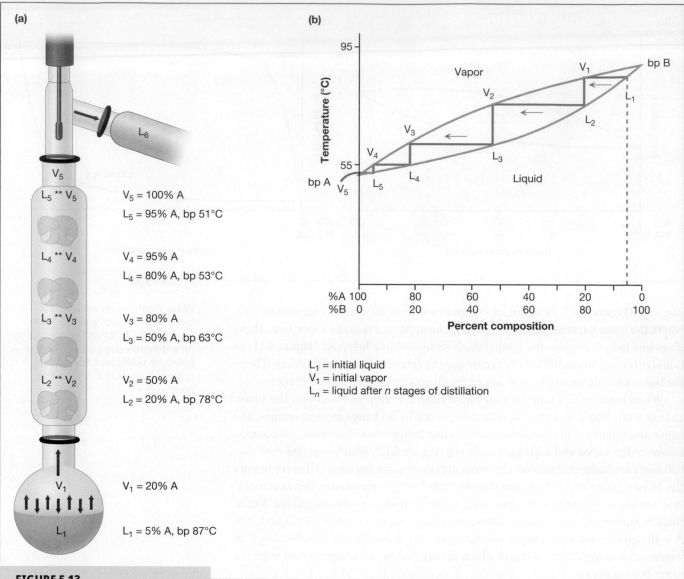

$V_5 = 100\%$ A

$L_5 = 95\%$ A, bp 51°C

$V_4 = 95\%$ A

$L_4 = 80\%$ A, bp 53°C

$V_3 = 80\%$ A

$L_3 = 50\%$ A, bp 63°C

$V_2 = 50\%$ A

$L_2 = 20\%$ A, bp 78°C

$V_1 = 20\%$ A

$L_1 = 5\%$ A, bp 87°C

(b)

L_1 = initial liquid
V_1 = initial vapor
L_n = liquid after n stages of distillation

FIGURE 5.13

(a) Fractional distillation of a simple two-component liquid mixture of initial composition 5% A and 95% B, where the boiling point of A is less than the boiling point of B. (b) The effect of fractional distillation is illustrated on the phase diagram.

▶ **Fractional Distillation**

5.3C | Vacuum Distillation

Vacuum distillation is particularly appropriate for compounds with very high boiling points, or for those that decompose at high temperature. The boiling point of a liquid is the temperature at which the vapor pressure of that liquid is equal to the applied pressure. So, the vacuum reduces the applied pressure within the simple or fractional distillation apparatus, thereby reducing the temperature at which the liquid boils. To accomplish this, the distillation apparatus is sealed from the atmosphere and connected to a vacuum pump. The distillation is carried out in the same way, except some care is needed to change flasks and collect fractions without disrupting the vacuum. A stopcock placed between the vacuum adapter and the receiving flask allows the distillation to proceed under vacuum while the receiving flask is removed and replaced.

The joints between the components of the apparatus need to be tightly sealed before vacuum is applied. The ground glass surfaces usually seal well enough on their own, but occasionally two joints will not match well enough for a good seal. Grease can be applied sparingly to help seal ground glass joint connections, but the grease can contaminate the desired products, it makes a big mess of the glassware, and it is difficult to clean up. Grease should only be used upon instruction from the TA or instructor. A thin strip of Teflon tape between the ground glass surfaces of the joint accomplishes the same objective with much less mess.

Boiling can be quite erratic in a vacuum distillation; occasionally the material in the boiling flask will "bump," or suddenly burst upward within the apparatus. Bumping can send the unpurified liquid mixture into the condenser and then into the collection flask. Use a boiling stick, boiling stone in the distillation flask, and/or rapid magnetic stirring to avoid this.

5.3D | Rotary Evaporation

Organic solvents are often used to dissolve reactants before carrying out reactions or separations in the organic lab. The solvents must subsequently be removed to recover the organic compounds that are in solution. Solvents having a low boiling point (usually 80°C or below) are commonly used to facilitate separation by vacuum distillation. There is a specialized device for this type of process called a rotary evaporator, or "rotovap" (**Figure 5.14**), that rotates a round-bottom flask during distillation, increasing the surface area of the solvent for more rapid solvent evaporation. The solvent evaporation causes the flask to get cold, so often a water bath is used to keep the flask at a constant temperature.

 Rotary Evaporation

By reducing the pressure, the boiling points of common organic solvents can be lowered to room temperature or below. This means that not much heat is required to rapidly boil away the solvent, leaving behind any solutes. Reducing the pressure and thus the amount of heat required helps to avoid damaging heat-sensitive organic compounds.

CAUTION: Before operating the rotary evaporator, check it carefully to make sure its glass parts have no cracks. Your sample should be in a round-bottom flask that is also free of any cracks (replace if necessary). Cracked glassware can break under vacuum, causing glass pieces to fly. This is one reason why you should always wear eye protection in the lab.

To operate the rotary evaporator, the flask containing organic solvent is affixed to the ground glass joint that points down toward the water bath. Hold the flask with your

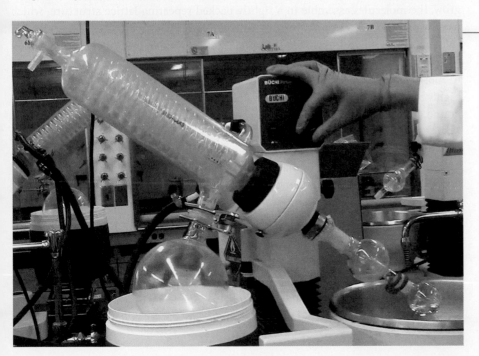

FIGURE 5.14

Rotary evaporator.
Courtesy of Gregory K. Friestad.

hand under it until you have securely clamped the ground glass connection together. Turn on the rotation of the flask, turn on the vacuum, and close the venting valve. Finally, lower the flask so that it just touches the water bath. Solvent will be removed. After the sample reaches a constant volume, reverse the order of steps to remove your flask: Open the venting valve, turn off the vacuum, raise the flask out of the water, stop the rotation, and place your hand beneath the flask while unclamping it.

<table>
<tr><td>5.3E</td><td>Steam Distillation</td></tr>
</table>

5.3E Steam Distillation

Steam distillation is used to purify organic compounds that are immiscible with water by distilling them along with water. Two immiscible liquids both contribute to the vapor pressure in the distilling flask, and as the water distills, the vapor phase (steam) carries some of the organic compound along with it. After the condensate is collected, the water can be separated from the organic compound because the two are immiscible.

A main reason for using this approach is to lower the temperature at which the organic compound distills. This not only makes it a more energy-efficient process, but also is more likely to avoid degradation of heat-sensitive organic functional groups.

The water may provide other beneficial effects. For example, boiling plant material along with water may disrupt cell walls, helping to release an organic compound from the matrix and allow it to distill more efficiently. This is a procedure commonly used in separating essential oils from various plants.

5.4 RECRYSTALLIZATION OF ORGANIC SOLIDS

If impurities are present in organic solids, the organic solid may be purified by recrystallization from a solution. When the amount of solute in solution exceeds the solubility, the solution is "supersaturated," and the solute will come out of solution. If you are fortunate, the solute will form crystals as it comes out of solution. As crystals grow, the molecules assemble in a tightly packed repeating lattice structure, which usually accommodates no impurities. Therefore, as a crystalline solid forms from a solution, impurities are excluded from the crystals. A rapid precipitation of powdery or amorphous solids may not exclude impurities, so a slow crystallization is preferred in most cases. Higher purity is usually observed when the solid forms crystals of highly regular geometric shapes or "crystal habits" that are characteristic to the compound. These may appear as needles, cubes, columns, or other geometries.

In the standard technique, crystallization is induced by lowering the temperature of a saturated solution. The compound is mixed with the minimum amount of hot solvent, just enough to dissolve the compound. Then the solution is allowed to cool, so that it becomes supersaturated. The crystals that form are recovered by vacuum filtration. The steps are summarized in more detail below.

 Recrystallization

5.4A Summary of the Basic Steps of Recrystallization

For most routine work, the following sequence of steps will generally be effective. Further details and more specialized techniques are described later.

1. Choose a solvent. The key is to identify one in which the solute has high solubility when hot, and low solubility when cold.

2. Dissolve the substance. First, heat the solvent in a separate Erlenmeyer flask. To an Erlenmeyer flask containing the substance, slowly add the minimum amount of hot solvent needed to dissolve the desired substance, and no more.

3. Optional hot filtration. If the desired substance is dissolved, but the impurities are not, remove the impurities via gravity filtration (see Figure 5.1) while keeping the solution and filtrate hot.

4. Cool slowly. Remove the solution from heat and allow it to stand undisturbed while crystallization proceeds. Here, patience is a virtue. After crystallization occurs, cool in an ice bath to ensure completion.

5. Recover the crystals. Use vacuum filtration (see Figure 5.2) to separate the crystals from the liquid filtrate (the **mother liquor**).

6. Optional repeat. Some of the desired substance may remain in the mother liquor. Evaporate some or all of the solvent and repeat the sequence to obtain a second batch (crop) of crystals.

<< **mother liquor**
The liquid phase that remains after removal of crystalline solid, e.g., by filtration.

5.4B | Choosing a Recrystallization Solvent

The fundamental requirement is that the compound is soluble at high temperature, and mostly insoluble at low temperature. In that scenario, a hot saturated solution can be cooled, forcing the solute to come out of solution. There is no single ideal recrystallization solvent to apply for all situations, and sometimes trial and error is required.

Several solvents are listed in **Table 5.1**, with boiling points and hazard notes. Except for dichloromethane and water, all of them present some flammability hazards, and should be kept away from flames. Diethyl ether, dichloromethane, and benzene introduce additional risks, and should generally be avoided for recrystallizations in the instructional lab except with special instructor permission.

TABLE 5.1

Some Solvents Suitable for Recrystallization

SOLVENT	bp (°C)	HAZARD NOTES
Diethyl ether	35	Extremely flammable
Dichloromethane (CH_2Cl_2)	40	Possible carcinogen
tert-Butyl methyl ether (MTBE)	55	
Acetone	56	
Hexane	69	
Ethyl acetate	77	
Ethanol	78	
Benzene	80	Carcinogen
Water	100	

To choose a good solvent, place 5–10 mg of the solid to be recrystallized into each of several vials, and add a few drops of various solvent candidates. An ideal solvent will not dissolve the solid when cold, but will dissolve most or all of it when heated. If the solid dissolves immediately in the cold solvent, reject it. If the solid doesn't dissolve at all, even at the boiling point of the solvent, reject it as well.

5.4C Recrystallization from Two-Solvent Systems

Two-solvent systems can give you much better control over solubility. The organic solid should have a high solubility in one of the solvents (the "good" solvent), and little or no solubility in the other (the "poor" solvent). Poor solvents can be either too polar (water) or not polar enough (hexane) to dissolve the organic solid. Once a pair is chosen, the polarity of the two-solvent system can be modified simply by varying the ratios of the two solvents. To enable this, the good and poor solvents must be miscible (mutually soluble with each other at all proportions). It is often helpful to choose a good solvent that has a lower boiling point than the poor solvent, so that the good solvent can be evaporated if too much has been added. Although the labels of "good" and "poor" may depend on the type of compound being recrystallized, **Table 5.2** lists some common two-solvent systems that are often useful for recrystallization of typical organic compounds.

In a typical two-solvent procedure, a small amount of the poor solvent is added to the compound, and the mixture is heated on a hot plate. Occasionally the compound dissolves in the poor solvent; if this occurs, simply cool and recrystallize as described previously (see section 5.4A) for the single-solvent method. If the compound does *not* dissolve in the poor solvent, then slowly add the good solvent, while keeping the mixture warm, until the compound is just dissolved. (If most of the material dissolves readily, leaving a persistently insoluble precipitate or cloudiness, remove these insoluble impurities via hot gravity filtration.) Remove the homogeneous solution from the heat and allow it to cool, so that it becomes supersaturated. If crystallization does not occur, cool the flask further in an ice/water bath. (Additional advice on inducing crystallization is given below.) Recover the crystals by vacuum filtration. A cold mixture of the two solvents may be used to wash the crystals, using a lower ratio of good solvent to avoid dissolving the crystals.

TABLE 5.2

Two-Solvent Systems for Recrystallization

GOOD SOLVENT	POOR SOLVENT
Ethanol (bp 78°C)	Water (bp 100°C)
Acetone (bp 56°C)	Water (bp 100°C)
tert-Butyl methyl ether (bp 55°C)	Hexane (bp 69°C)
Dichloromethane (bp 40°C)	Hexane (bp 69°C)
Ethyl acetate (bp 77°C)	Hexane (bp 69°C)

5.4D Inducing Crystallization

Crystals don't always form spontaneously; you may need to intervene to induce crystallization. This process is a combination of skill, art, and luck. There are various ways to induce crystallization, but the basic process is to prepare a saturated solution of the organic solid, then change the conditions in some way that lowers the solubility. This is usually accomplished by lowering the temperature (because the solubilities of solids decrease with decreasing temperature), causing the solution to be supersaturated—that is, the amount of material in solution is greater than its solubility will allow—so that the solid must come out of solution. If crystals don't grow spontaneously, crystallization can be encouraged by introducing **nucleation sites** where crystal growth can initiate. The best way to do this is by "seeding" the supersaturated solution with a trace amount (< 1 mg—not enough to affect the yield calculation) of the same compound that has been previously crystallized. Sometimes, the seed crystals initiate a rapid crystal growth, which can be strikingly beautiful to

nucleation site >>
The origin of crystal growth from solution, where molecules begin to organize to form a crystalline solid.

watch. If seed crystals are unavailable, crystallization can sometimes be induced by scratching the inside of the Erlenmeyer flask with a glass rod or pipet, which creates nucleation sites on the surface of the glass.

The best purification occurs when crystals are allowed to grow slowly from a homogeneous solution over a period of time, usually from a few minutes to a few hours. Forcing a compound out of solution by changing the solubility too rapidly leads to a liquid or amorphous solid precipitate rather than crystals. A precipitate is likely to trap more impurities within the solid, thus giving an unsatisfactory purification.

5.4E | Recovering the Crystals

After crystallization has occurred, most of the impurities are generally left in the liquid phase. The two-phase mixture can then be separated using vacuum filtration to recover the crystals. The liquid filtrate, called the mother liquor, contains soluble impurities, along with additional desired compound that had yet to be crystallized. A second or third batch (or crop) of crystals can sometimes be obtained by evaporating some of the solvent from the mother liquor and then cooling it again. Thin-layer chromatography (section 5.5) can be used to assess whether the mother liquor still contains enough of the desired compound to make it worth pursuing an additional crop of crystals.

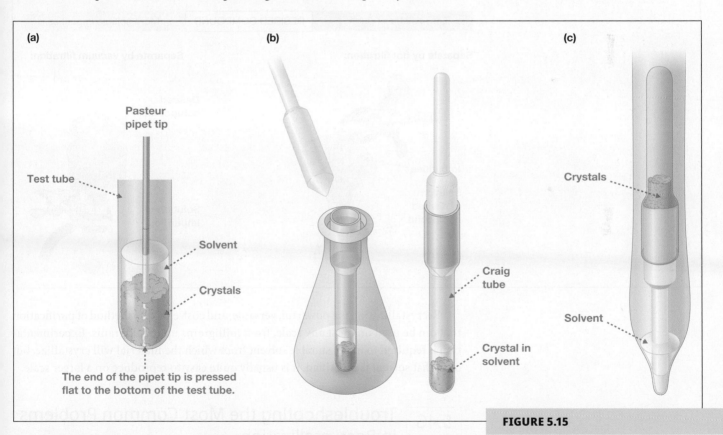

(a)

Pasteur pipet tip

Test tube

Solvent

Crystals

The end of the pipet tip is pressed flat to the bottom of the test tube.

(b)

(c)

Craig tube

Crystal in solvent

Crystals

Solvent

FIGURE 5.15

(a) Pipet tip pressed to bottom of test tube for microscale filtration. (b) Craig tube for separating crystals from solvent with (c) the aid of a centrifuge.

On small scales (e.g., < 50 mg), recrystallizations are sometimes more conveniently performed in a test tube, and the solvent may be removed with a Pasteur pipet. Holding the tip of the pipet firmly against the bottom of the tube creates a very small gap that will not let larger crystals pass as the solvent is drawn up into the pipet (**Figure 5.15a**).

A more elegant test tube recrystallization uses a Craig tube (**Figure 5.15b**). This tube has a smaller diameter restriction above the crystallization mixture, and a glass insert that nests into it. The glass pieces are machined so that they fit tightly together and only solvent can pass. Once assembled with the crystallized mixture inside, the

Craig tube is inverted inside a larger test tube (**Figure 5.15c**) that is placed in a centrifuge. Centrifugal force pushes the liquid past the insert and the liquid collects in the larger tube, leaving the crystals behind in the Craig tube.

5.4F   Hot Filtration of Less Soluble Materials

Depending on how the desired compound is formed or isolated, there may also be impurities that are less soluble than the desired compound. When recovering the compound by vacuum filtration, these less soluble impurities will not be separated. Less soluble impurities may be detected as a persistent cloudiness that does not clear up upon adding more solvent. These insoluble materials may be removed by hot filtration or gravity filtration prior to crystallization (**Figure 5.16**). After the hot filtration, the filtrate contains the desired compound. Crystallization may then be induced, and finally, the desired compound may be isolated by vacuum filtration.

FIGURE 5.16

Different types of filtration and their purposes during a recrystallization.

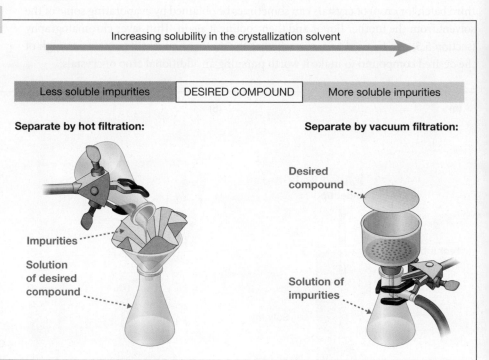

Recrystallization is a powerful, versatile, and cost-effective method of purification that can be used on most any scale, from milligrams up to kilograms. Experimentation is required to find a suitable solvent from which the material will crystallize, but once that solvent is identified, it is usually quite easy to reproduce on a larger scale.

5.4G   Troubleshooting the Most Common Problems in Recrystallization

PRECIPITATION

If the solid comes out of solution quickly, this is referred to as precipitation, and some of the impurities are often trapped within a solid precipitate. Such solids may resemble a crust or powder with no visible crystallinity, and are said to be amorphous. The material should be re-dissolved by heating, or by adding more of the good solvent, if a two-solvent system is being used, and then allowed to cool more slowly.

OILING

If the material comes out of solution at a temperature above its melting point, it will precipitate as an oil. Cooling this two-phase mixture may result in solidification rather than a proper crystallization, and the solid that results likely contains significant amounts of impurities. A small additional portion of solvent should be added, and the mixture reheated to form a homogeneous solution. With more solvent present, the hot solution may cool below the compound's melting point before it becomes supersaturated, so that it comes out of solution as a solid rather than as an oil.

NOTHING COMES OUT OF SOLUTION

If neither crystals nor oil are forming, there's probably too much solvent, or if a two-solvent system is being used, there may be too much of the good solvent. Excess solvent can be removed by using a rotary evaporator, or by adding a boiling chip and placing the Erlenmeyer flask back on the hot plate. If a two-solvent system is being used, and if the good solvent has a *lower* boiling point than the poor solvent, then boiling will help because the good solvent will be more rapidly evaporated. If the good solvent has a *higher* boiling point, then it will make the problem worse. In this case, add more of the poor solvent while heating, then allow the solution to cool again.

5.4H │ Specialized Techniques for Recrystallization

Occasionally a compound can be best crystallized by very slowly changing the composition of the solvent. This can be achieved by slow diffusion of the poor solvent into the good solvent, which may take hours or days. The methods described here are not commonly used in the introductory organic experiments, but they can be very handy in the event that typical recrystallization techniques fail.

SOLVENT LAYERING

The solvent layering technique requires two miscible solvents, where the good solvent is more dense than the poor solvent. A solution of the compound in the good solvent is placed in a test tube, and the poor solvent is slowly added by pipet at the surface, avoiding mixing, so that the less dense poor solvent rests on top as a separate layer. The test tube is then stoppered and allowed to stand undisturbed. Over time (e.g., waiting until the next lab period), the solvents will diffuse across the interface of the layers. As more of the poor solvent mixes into the lower layer of good solvent, the solubility of the compound slowly decreases, leading to crystallization.

SOLVENT VAPOR DIFFUSION

Figure 5.17 shows how a poor solvent can be made to diffuse via its vapor phase into a solution of the compound in the good solvent. A small, open container of the compound dissolved in a minimum amount of the good solvent is placed inside a larger container of the poor solvent. The larger container must have an air-tight seal to prevent the vapors from escaping. The two solvents will both be in equilibrium between their liquid and gas phases, and over a few hours or days, will diffuse into one another, reducing the solubility of the compound and causing it to crystallize.

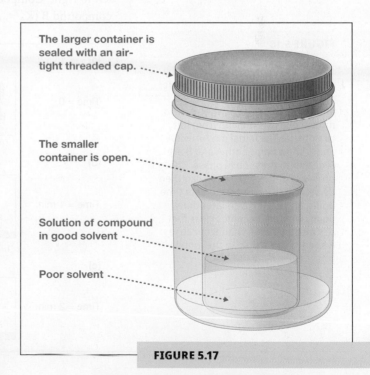

The larger container is sealed with an air-tight threaded cap.

The smaller container is open.

Solution of compound in good solvent

Poor solvent

FIGURE 5.17

Simple apparatus for solvent vapor diffusion during recrystallization.

5.5A Background

Chromatography is one of the most ubiquitous methods of analyzing and purifying organic compounds from mixtures. This technique, originally used to separate plant pigments, encompasses a variety of sophisticated methodologies that allow for the separation, isolation, and identification of the components of a mixture. While there are many types of chromatography, the fundamental basis for this technique is the distribution of the individual components of a mixture between two phases: the **stationary phase and the mobile phase**. For any given compound (A), there is a rapid equilibrium between phases; the compound spends some time adsorbed on the stationary phase and some time dissolved in the mobile phase.

stationary phase and mobile phase >>
In chromatography, the stationary phase is a motionless material that accompanies another phase that is moving (the mobile phase). Compounds differ in how strongly they associate with the stationary phase, causing them to be separated as they travel at different rates along with the mobile phase.

$$A_{(mobile)} \xrightleftharpoons{K} A_{(stationary)} \qquad K = \frac{[A_{(stationary)}]}{[A_{(mobile)}]}$$

The equilibrium constant K depends upon the intermolecular attractions that the compound experiences with both the stationary and mobile phases. If there is a weak intermolecular attraction to the stationary phase, the compound will be mostly in the mobile phase ($K < 1$), and will travel rapidly along with the mobile phase. Conversely, if the compound has strong intermolecular attractions with the stationary phase, the compound will travel much more slowly because it spends most of its time in the stationary phase ($K > 1$).

Figure 5.18 depicts a cross section of a chromatographic separation at three different time points, with mobile and stationary phases. The mobile phase is moving left to right. Compound A has a weaker association with the stationary phase than compound B ($K_A < K_B$). As a result, less of A is adsorbed on the stationary phase at

FIGURE 5.18

Cross section of a chromatography column, showing chromatographic separation of two components A and B from a mixture, where the mobile phase is moving from left to right. (a) The initial mixture of A and B shows that both A and B are together in the mobile phase. The progress of the separation can then be seen at (b), an intermediate time point, and (c), an even later time point, where it becomes clear that A is moving faster than B because A spends more time in the mobile phase ($K_A < K_B$).

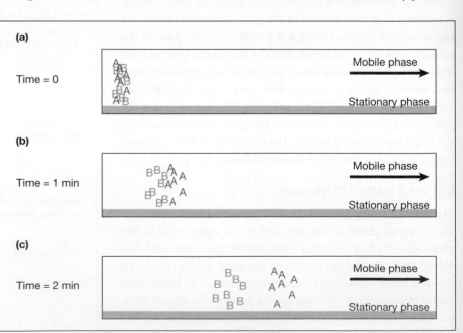

any given time. As the mobile phase moves from left to right, B moves more slowly than A because a greater proportion of B is in the stationary phase. Alternatively, A moves faster than B because it "spends more time" in the mobile phase.

5.5B Gas Chromatography

Gas chromatography (GC) is a common analytical technique used to separate volatile organic compounds. It may be applied to identify the compounds within an unknown mixture (qualitative analysis) or to determine their relative amounts (quantitative analysis).

In a typical gas chromatography instrument (**Figure 5.19**), a long tube called a GC column is placed in a temperature-controlled oven. The oven includes both a heating element and a cooling fan, so that temperature can be accurately adjusted either up or down. The GC column is where the separation takes place. It is a stainless steel or glass capillary tube, 2 m or more in length, coated on its interior surface with a stationary phase. The mobile phase is an inert gas (usually helium), also called a carrier gas, which is passed through the column at a controlled **flow rate**. A small amount (e.g., 1 μL) of a liquid or gaseous sample is injected into the column. The compounds in the sample are carried along by the mobile phase and detected as they emerge from the outlet.

A flame ionization detector (FID) is commonly used with GC. The outflow from the GC column passes through a hydrogen–air flame, and when an organic compound is present, ions are produced and attracted by a voltage in the detector. The resulting signal is plotted versus time to produce a chromatogram, with peaks appearing at specific times that are characteristic of the compounds present.

As in other types of chromatography, compounds travel through the column at different rates because they exist in equilibrium between the stationary and mobile phases, with different equilibrium constants for association with the two phases. They spend some time adsorbed ("stuck") on the stationary phase as a liquid, and some time moving with the carrier gas as a vapor. Compounds that associate more strongly with the stationary phase take longer to pass through the column. Compounds with higher vapor pressure are more associated with the mobile phase. Consequently, *boiling point* is the most important property for separation via GC.

Polarity, however, may also affect GC behavior. If two compounds have similar boiling points but very different polarities, separation can still occur. For many GC columns the *less* polar one will emerge from the column first. Solid samples can be run on the GC by first dissolving them in an appropriate solvent. Care should be taken to inject only volatile solids; if the solid cannot be vaporized at the temperatures of the injection port and oven, it may clog the system or damage the column.

<< flow rate
In chromatography, the rate of input and output of a mobile phase.

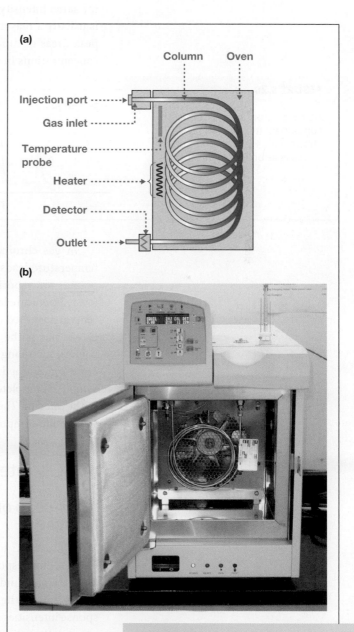

FIGURE 5.19

(a) Schematic diagram of a gas chromatograph. (b) A typical gas chromatograph, with the oven door open to reveal the column and cooling fan. The oven door is closed during operation.

(b) Courtesy of Gregory K. Friestad.

A plot of GC detector response versus time is called a gas chromatogram (**Figure 5.20**). Each peak in the chromatogram has a specific **retention time**, which is the amount of time between when the sample was injected and when it emerged from the column.

The detector response is proportional to the amount of compound passing through it, so the area under a peak is proportional to the total amount of compound in the sample. The ratio of peak areas in a single chromatogram is equal to the ratio of compounds in the mixture, as long as we assume the compounds give the same intensity of response to the detector. In the introductory organic lab, this is usually a reasonable assumption. For more precise measurements, however, the peak areas of different components must be compared with standards of defined concentrations in order to account for the different detector responses.

FIGURE 5.20

GC chromatogram of a three-component mixture. The table shows that retention times often increase as boiling points increase.

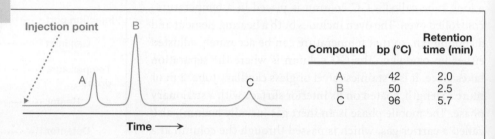

Compound	bp (°C)	Retention time (min)
A	42	2.0
B	50	2.5
C	96	5.7

In gas chromatography, the oven temperature affects retention times—high temperature leads to a short retention time and little separation because all compounds are vaporized and thus move at the same rate as the mobile phase. A very low temperature leads to impractically long retention times because the compounds remain adsorbed on the stationary phase. In addition, diffusion causes the peaks to spread out as the retention time increases, so compounds that are retained in the column for a long time may appear as broad, ill-defined peaks. The injector is generally maintained at a much higher temperature than the column to ensure that the sample is completely vaporized before it reaches the column and does not condense in the injector. The temperature of the detector is also set to prevent condensation of the sample components.

5.5C Interpreting GC Data

USING GC TO QUANTIFY THE RATIOS OF COMPONENTS OF A SAMPLE MIXTURE

We can use GC to quantitatively evaluate a mixed sample of known components. The detector response is proportional to the amount of compound passing through it, so the area under a peak (the integration) is proportional to the total amount of compound in the sample. Assuming two compounds A and B give the same response intensities at the detector, the ratio of peak areas in a single chromatogram is equal to the ratio of compounds A and B in the mixture. Peak areas for A and B are calculated using the following equation:

$$\text{Area} = (\text{peak height})(\text{peak width at } 1/2 \text{ height})$$

Using this method, the ratio in a two-component mixture can be found by simply dividing the area of the larger peak by the area of the smaller one. If there are more components, the smallest of the peaks is identified, and the relative amount of each component is calculated by dividing the area of each larger peak by the smallest peak.

Worked Example

Compounds A and B have equal detector responses. For the sample chromatogram shown in **Figure 5.21**, area$_B$/area$_A$ ≈ 2.5, so the ratio of component B to component A in this sample is 2.5 to 1.

This area calculation is an effective but "low tech" way of integrating the areas under the peaks. Modern GCs can integrate peaks very accurately, and typically the integration data for each peak are included in a table along with the printed chromatogram.

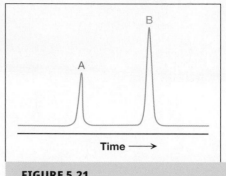

FIGURE 5.21

USING GC TO DETERMINE PURITY

When evaluating the purity of a compound by GC, the observation of a single, large peak suggests there is only one component in the mixture. However, more than one component may have the same retention time. For example, in **Figure 5.22**, the peak on the left shows a sample that appears to be pure. The chromatogram on the right shows a sample that is impure. To confirm that a peak consists of just one component, additional evidence may be needed, such as using a different GC column, or a variety of conditions (different temperatures, flow rates, etc.). If a single peak is still seen in all these analyses, this strengthens the evidence that the sample is pure.

FIGURE 5.22

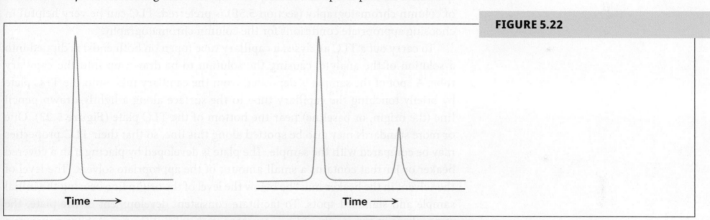

USING GC TO IDENTIFY AN UNKNOWN

Retention times in GC can help to identify unknown compounds. One method is to compare the retention time of the unknown with those of authentic samples of known compounds. The presence of a peak does not conclusively identify the structure of the compound, so usually further information about an unknown is needed to establish its identity.

If there are only a couple of possibilities for the identity of the unknown, and authentic samples are available, then you can add a known compound to the sample (this technique is called "spiking"). If the "spiked" known compound matches the unknown, then one peak will increase in size. If a new peak appears, or if the one peak develops a shoulder, then the "spiked" compound doesn't match the unknown.

Additional information about each peak may be obtained with an instrument that routes the GC output directly into a mass spectrometer (GC-MS). As the components come out of the end of the GC along with the mobile phase, they are routed into the mass spectrometer, and a mass spectrum is obtained from the components of each separate peak in the chromatogram. Mass spectrometry is discussed in detail in Chapter 8.

5.5D | Thin-Layer Chromatography

Thin-layer chromatography (TLC) is a separation technique that is used to determine the purity of a compound, the status of an ongoing reaction, or as a preliminary means of identification. The basic principles discussed for GC chromatography apply here as well—namely, there is a stationary phase and a mobile phase, and the analytes travel with the mobile phase at different rates depending on how strongly they associate with the stationary phase. In TLC, the stationary phase is the TLC plate, a thin layer of finely powdered silica gel (SiO_2) or alumina (Al_2O_3) that is affixed to a glass slide or to a thin sheet of aluminum or plastic. The SiO_2 or Al_2O_3 of the stationary phase is a polar solid to which the components of a mixture may adsorb with different affinities depending on their polarities. The mobile phase may be a single organic solvent or a mixture of a nonpolar organic solvent, such as hexane or petroleum ether, plus varying amounts of a more polar solvent, such as ethyl acetate, to adjust the polarity of the solvent mixture. As the mobile phase travels through the stationary phase, the components of a mixture are carried along at different rates because of their different affinities for the stationary phase, resulting in separation.

Like gas chromatography, TLC is used primarily as an analytical technique, because the amount of material loaded onto the TLC plate is generally very small and often not worth recovering afterward. When the materials must be recovered in quantities that are useful for subsequent experiments, the closely related technique of column chromatography (section 5.5F) is preferred. TLC can be very helpful in choosing appropriate conditions for the column chromatography.

To carry out a TLC analysis, a capillary tube (open on both ends) is dipped into a solution of the analyte, causing the solution to be drawn up into the capillary tube. A spot of the sample is deposited from the capillary tube onto the TLC plate by briefly touching the capillary tube to the surface along a lightly-drawn pencil line (the origin, or baseline) near the bottom of the TLC plate (**Figure 5.23**). One or more standards may also be spotted along this line, so that their TLC properties may be compared with the sample. The plate is developed by placing it in a covered beaker or jar that contains a small amount of the appropriate solvent. The level of the solvent in the beaker must be below the level of the origin line bearing the initial sample and standard spots. To facilitate consistent development of the plate, the atmosphere in the jar should be saturated with solvent vapors. A filter paper or a small section of a paper towel is used to help keep the atmosphere in the container saturated, but the TLC plate should not touch the filter paper.

Capillary action draws the solvent up the plate, and the leading edge of the solvent can usually be observed visually as it travels. This is called the **solvent front**. When the solvent front is near the top of the plate, the development is complete, so the plate is removed from the beaker and the solvent front is immediately marked with a pencil before the mobile phase evaporates.

The location of each spot is then noted. Plates that contain a UV fluorescent material facilitate this. If the spots lack color for visualization, they can be visualized using an ultraviolet lamp, if the compound absorbs UV light. For some compounds, staining the plate with a chemical stain is preferred. A simple chemical stain entails placing the TLC plate into a closed jar containing a few crystals of iodine; the iodine vapor reacts to give a color at the location of spots on the plate. More widely effective is dipping the plate into a dilute aqueous $KMnO_4$ solution, then heating it on a hot plate or with a heat gun. The plate will appear purple, except for yellow spots where compounds are located. Other dip-and-heat stains include anisaldehyde, 2,4-dinitrophenylhydrazine, and phosphomolybdic acid.

After developing the plate, an initial spot containing more than one component should now show multiple spots that traveled different distances from the origin.

solvent front >>
In thin-layer chromatography, the distance traveled by the solvent during development of a plate.

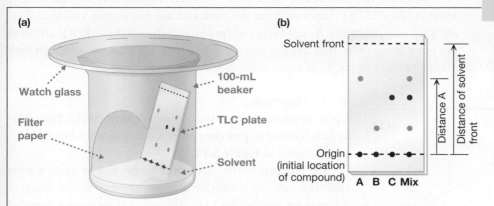

FIGURE 5.23

(a)

Watch glass

Filter paper

100-mL beaker

TLC plate

Solvent

(b)

Solvent front

Distance A

Distance of solvent front

Origin (initial location of compound)

A B C Mix

The components can be analyzed by determining the distance each spot traveled and comparing it with standards. For each spot on the TLC plate, a characteristic value R_f or "ratio to the front" (also sometimes called "retention factor") can be calculated. For example, for the yellow spot A (Figure 5.23b), the R_f is calculated as follows:

$$R_f = \frac{\text{distance A from origin}}{\text{distance of solvent front from origin}}$$

The R_f value is defined as the ratio of the distance traveled by a spot (measured from the center of the spot) to the distance traveled by the solvent (Figure 5.23). The R_f value is characteristic for a given compound as long as the polarities of the stationary phase and mobile phase are carefully controlled. This level of control is difficult to achieve, however, so there are no tables of R_f values in the chemical literature. Instead, standards of the compounds known to be in the mixture are included as separate spots on the plate (Figure 5.23), so that each standard R_f value can be matched with the components of a mixture on the same plate, ensuring the same conditions for the analysis.

The difference in R_f values between two compounds will also vary with the solvent, although generally the R_f values appear in the same order (i.e., the higher R_f compound will still be the highest). The choice of developing solvent is crucial. With a solvent that is too polar, all of the spots will run to the top of the plate, and there will be no difference in R_f. With a very nonpolar solvent, on the other hand, the spots will not move from the baseline, and again there will be no difference in R_f. Generally, the best separations are achieved by selecting a solvent that moves the spots to the middle areas of the plate—namely, $R_f = 0.3$–0.7. To simplify the selection of solvent, mixtures of polar and nonpolar solvents can be used in various ratios to adapt the solvent polarity to the polarities of the compounds in the mixture. A commonly used solvent pair is ethyl acetate and hexane.

▶ **Thin-Layer Chromatography**

5.5E | Interpreting TLC Data

USING TLC TO EVALUATE THE PROGRESS OF A REACTION

When evaluating a reaction mixture, the disappearance of the spot representing starting material and the appearance of a new spot over time indicate that the original compound has been converted to something else. This means that the reaction is proceeding or has gone to completion. Generally speaking, the more polar a compound, the lower its R_f value, and vice versa. Thus, for a given set of conditions, the R_f values of two spots on a TLC plate may provide some evidence of the identity of a compound, and the success (or failure) of a reaction.

TLC is convenient to carry out directly on reaction mixtures because volatile solvents like CH_3OH evaporate from the plate and are not usually visible in TLC analysis. Also, inorganic by-products and reagents (such as NaCl or KOH) are often too polar to move, so they usually don't interfere with the analysis, although they may sometimes be visible as a spot at the origin.

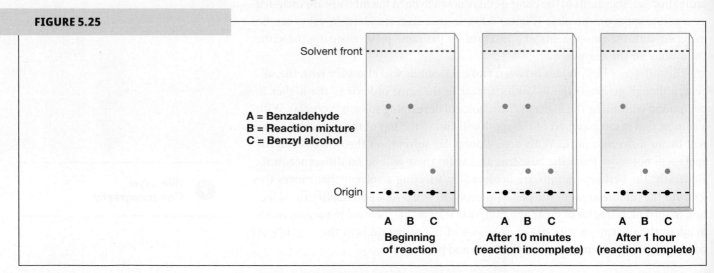

FIGURE 5.24

Benzaldehyde

NaBH₄
CH₃OH

Benzyl alcohol

Worked Example

Upon treatment with sodium borohydride ($NaBH_4$), benzaldehyde is reduced to give the corresponding alcohol, benzyl alcohol, as shown in **Figure 5.24**.

The progress of the reaction is monitored by using a glass capillary tube to withdraw a trace amount of the reaction mixture for analysis by TLC. The reaction mixture is sampled before the reaction begins, after 10 minutes, and again after 1 hour, and the analysis is carried out on three different TLC plates (**Figure 5.25**). In each of the three TLC plates, authentic standards are placed on the plate for comparison; spots A and C are authentic samples of benzaldehyde and benzyl alcohol, respectively.

As the plates are developed, the spots move vertically from the origin in a lane. The spots in lane B (the reaction mixture) are compared with lanes A and C (the standards) to determine if the reaction is complete. After 30 minutes, the reaction is incomplete, because lane B contains both benzaldehyde and benzyl alcohol. It is successfully completed after 1 hour, however, because the higher R_f benzaldehyde spot is no longer present in lane B. The relative R_f values in this case are as expected—that is, the aldehyde has a higher R_f value than the alcohol, because the alcohol is more polar (it can serve as both a hydrogen bond donor and a hydrogen bond acceptor).

FIGURE 5.25

Solvent front

A = Benzaldehyde
B = Reaction mixture
C = Benzyl alcohol

Origin

A B C
Beginning
of reaction

A B C
After 10 minutes
(reaction incomplete)

A B C
After 1 hour
(reaction complete)

USING TLC TO IDENTIFY COMPOUNDS IN MORE COMPLEX MIXTURES

More complex mixtures can also be analyzed by TLC, because the procedure used for the two-component mixture of benzaldehyde and benzyl alcohol can be extended to several components. That is, the analyte is placed in one lane, and an extra lane is added for each of the authentic standards to be compared with the analyte, using a wider TLC plate if necessary to fit additional lanes. The presence or absence of the various components can be determined by the presence or absence of a spot in the analyte lane at the R_f of the standard.

Worked Example

Nutritional supplements may contain components that may or may not be beneficial, so it is important to determine which ingredients are present or absent from a particular commercial product. The nutritional supplement is placed in one lane, and each authentic standard is placed in its own lane. After developing, the absence of a spot in the analyte lane at the R_f of an authentic standard shows that that component is absent (or below the level of detection). If that spot is present at the matching R_f, however, it indicates the presence of that component. The presence of a spot is weaker evidence than the absence of a spot, though, because TLC alone is insufficient to prove the identity of the component. Many compounds may have the same R_f, so the spot could be from something else. Further evidence, such as IR and NMR spectra, may be needed in order to confirm its identity.

USING TLC TO INDICATE PURITY

A pure compound should produce a single spot in TLC. Two (or more) spots in a single lane indicate that the compound is impure. However, while TLC can often show clearly that a substance is impure, TLC may fail to detect an impurity sometimes, too. Impurities that are low in concentration, or unresponsive to UV light or chemical stain, may be present but not visible. Different compounds (and impurities) may exhibit very similar R_f values on TLC and thus may appear as a single spot, making it impossible to distinguish them using this method. As such, other experimental techniques should be used to confirm the purity of a substance, even if it appears to be a single compound by TLC.

Worked Example

Analysis of a sample by TLC shows the presence of two components, the desired compound, B, and a higher R_f impurity, A. After purification by column chromatography (section 5.5F), a series of fractions are obtained, and to find out which fractions contain A and which contain B, they are analyzed by TLC. A spot for each fraction is placed in separate locations along the baseline on the TLC plate (Figure 5.25). After developing the plate, the R_f values of spots from each fraction are compared with the standards to see if they match with A or B (or contain both). The desired compound can then be recovered from the fractions in which it has been observed by TLC.

5.5F Column Chromatography

Column chromatography involves a mobile phase of organic solvent that passes through a column of finely powdered solid material (often SiO_2 or Al_2O_3). The mobile phase carries the components of the mixture through the column at different rates, thus allowing them to separate. This separation is analogous to TLC, but can be used for purification of larger amounts of material. The method is especially important in cases where crystallization and distillation are unsuccessful. It is easiest to apply when the desired compound and its impurities have different R_f values in thin-layer chromatography (TLC), because it is governed by the same principles as TLC, except that the adsorbent or stationary phase is packed in a glass tube or column, rather than spread on a thin plate (**Figure 5.26**).

In column chromatography, the stationary phase of SiO_2 (silica gel) or Al_2O_3 (alumina) is a polar solid to which the components of a mixture may adsorb with different affinities, depending on their polarities. The mobile phase is commonly a nonpolar organic solvent, such as hexane or petroleum ether, plus varying amounts of a more polar solvent, such as ethyl acetate, to adjust its polarity.

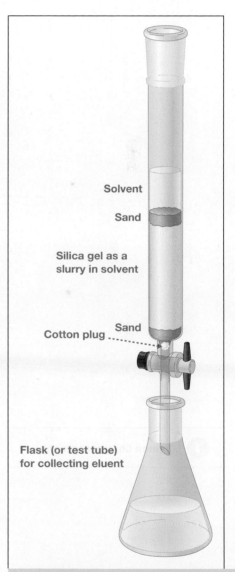

Solvent

Sand

Silica gel as a slurry in solvent

Sand

Cotton plug

Flask (or test tube) for collecting eluent

FIGURE 5.26

Setup for column chromatography, with elution in progress. Columns for this purpose are available in a variety of diameters and lengths, depending on the quantity of material to be separated.

More polar materials associate more strongly with the polar stationary phase, whereas nonpolar compounds associate weakly with the polar stationary phase. Because it is mostly in the mobile phase, a nonpolar compound is carried along more rapidly through the column. A polar compound spends most of the time bound to the stationary phase, so it moves more slowly. If the polarity of the mobile phase is increased, both components travel more rapidly through the column.

The sample mixture starts at the top of the column, and the mobile phase is passed through the column, moving the different components of the sample through the column at different rates. The components exit the column at different times, and can be collected in a number of fractions corresponding to materials that are less polar (early fractions) and more polar (later fractions).

5.5G | Typical Procedures for Column Chromatography

CHOOSING A SOLVENT

As with TLC, the choice of solvent system is crucial for good separation, and the best separation is often achieved by using solvent mixtures. As a rule of thumb, a good solvent system to start with is one in which the least polar component of the mixture has an R_f value of 0.3 in TLC. A single solvent mixture may be used to develop the column, or a solvent system that gradually increases in polarity (a polarity *gradient*) may be utilized. For example, a column may be developed starting with a low-polarity solvent, such as 10:1 hexane/ethyl acetate, and as fractions are collected, the developing solvent is changed from 10:1 to 5:1, then 3:1 hexane/ethyl acetate.

PACKING THE COLUMN

A cotton plug is loosely tamped into the constricted outlet tube at the bottom of the column. Some columns are made with a sintered glass frit to retain the column packing, and the cotton plug is unnecessary for columns of this type. A layer of sand is then added, and the top of the sand is made flat by gently tilting and tapping the side of the tube. The stationary phase is packed into the column, avoiding any gaps, bubbles, or cracks. The column may be packed "wet" by adding some solvent first and then pouring a solvent–adsorbent slurry into the tube, or "dry" by filling it with dry adsorbent and then adding the solvent. *CAUTION: Handle silica gel in the hood—fine particles of silica dust are harmful if inhaled.* It is important to keep the top and bottom surfaces of the adsorbent column as flat as possible—otherwise the separation will not be as effective. A protective layer of sand is placed on top to protect the top surface during the further addition of solvents.

 Column Chromatography

LOADING THE COLUMN

In rare cases, a solid mixture may be added directly. More commonly, the mixture to be separated is dissolved in a small amount of the chosen solvent [about three times the amount of the mixture, plus a few drops of dichloromethane (CH_2Cl_2), if needed to dissolve the mixture] and added carefully at the top of the column, so as not to disturb the packing. After this is allowed to drain down to the surface of the adsorbent, flow is stopped and an additional small portion of solvent is used to rinse the source flask, and this is also added at the top of the column. At this point there will be a visible band in the silica gel, usually white or yellowish, just below the sand. For good separation, this band must be as narrow as possible. Thus, avoid using excessive solvent to load the column because that will make this band wider.

DEVELOPING THE COLUMN

The column is developed, or *eluted*, by adding solvent to the top and collecting fractions of the *eluate* that comes out of the bottom. Gravity may be enough to elute the column at a reasonable rate. Often more efficient elution is obtained using "flash" chromatography, in which a slight air or nitrogen pressure is added to push the solvent through the column more rapidly. As the column is eluted, fractions may be collected in test tubes or small flasks, depending on the volume and the number of fractions. Add more solvent as needed—do not allow the solvent level to drain below the top surface of the column packing.

RECOVERING THE PURIFIED COMPOUND(S)

After the fractions are collected, they are analyzed by TLC to determine which fractions contain the compounds of interest (**Figure 5.27**). Ideally the compounds will be in pure form, as confirmed by a single spot on TLC. Fractions that contain the same single component may be combined, and the solvents can be removed by rotary evaporation to obtain the pure compound. Fractions with no detectable component at all may be discarded. Some fractions may contain mixtures, and they can be subjected to a repeat of the column chromatography if it is critical to recover every last milligram of material.

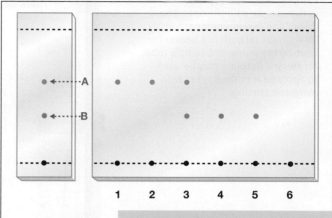

FIGURE 5.27

TLC plates of a mixture prior to column chromatography (*left*) and six fractions collected from the column (*right*). Fractions 1, 2, 4, and 5 show clean separation of the two pure components, while fraction 3 contains a mixture.

5.6 DETERMINATION OF PHYSICAL PROPERTIES

5.6A Boiling Point Determination

The boiling point (bp) of a liquid is the temperature at which the vapor pressure of the liquid equals the atmospheric pressure. It is a characteristic physical property of a compound, so experimental boiling point data can be compared with literature values to provide evidence about the identity of a compound.

The best way to determine the boiling point of a liquid is to distill it and record the temperature at which it comes over in the distillation. When determining the bp of an unknown liquid, however, where only a small amount of sample is available, a full distillation is out of the question. So, in order to determine the boiling point in such circumstances, a microscale technique has been developed to obtain boiling point data on liquid samples as small as 5–10 drops.

In a microscale boiling point determination, 0.5–1.0 mL of sample is placed in a small (75 × 150 mm) test tube that is attached to a thermometer via a rubber band (**Figure 5.28**). A capillary tube, with one end closed, is placed, open-end down, into the sample. The test tube/thermometer is then submerged in an oil bath. The oil bath is warmed slowly, until the open end of the capillary has a steady and rapid stream of bubbles emerging from it. The temperature is then gradually lowered while watching the stream of bubbles slow to a stop. When the last bubble emerges, the vapor pressure of the liquid equals the atmospheric pressure, and at that moment the temperature is recorded as the boiling point.

 Boiling Point Measurement

FIGURE 5.28

Experimental setup for microscale boiling point determination. (a) The sample tube, capillary tube placed in the sample with the open end down, and thermometer assembly. (b) A simple oil bath in a beaker for heating the immersed sample tube. An electric hotplate may be used to apply heat to the beaker with magnetic stirring.

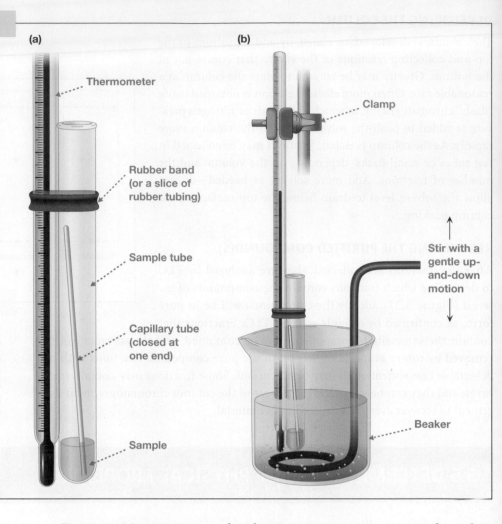

(a)

Thermometer

Rubber band (or a slice of rubber tubing)

Sample tube

Capillary tube (closed at one end)

Sample

(b)

Clamp

Stir with a gentle up-and-down motion

Beaker

decomposition >>
An undesired reaction that diminishes the amount of a desired compound.

Decomposition may occur when heating some organic compounds to their boiling point, especially while they are exposed to air. If the analysis is to be repeated, use a fresh sample, assuming sufficient material is available.

5.6B | Melting Point Determination

Melting points of crystalline solids are simple to obtain and are useful for identifying organic compounds and assessing their purity. A sample of the compound is placed in the closed end of a capillary tube by pressing the open end onto the solid (**Figure 5.29**), inverting it, and tapping it gently on the benchtop to move the crystals to the closed end. The tube should contain about 2–3 mm of solid. If the solid is difficult to get to the bottom, the tube can be dropped down a long tube held vertically on the benchtop.

The capillary containing the compound is slowly heated in an oil bath along with a thermometer, as in Figure 5.28, or in a heated metal block device such as a Mel-Temp (Figure 5.29). While observing the sample, the temperature is monitored with a thermometer or thermocouple. Heating at a rate of less than 5°C per minute, especially near the expected melting point, gives the best results. Two temperatures are recorded, one when melting first begins, and another at the point when melting is complete and the sample is a homogeneous liquid. Experimental melting point data are reported as a range of these two temperatures in the following format: 112–113°C.

 Melting Point Measurement

(a)

Melting point sampling:
1. Press the open end of the capillary tube into the solid.
2. Invert the capillary tube.
3. Tap the closed end of the capillary tube on the bench top to move the solid down into the closed end.

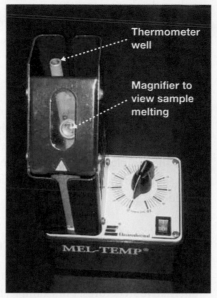

(b)

Thermometer well

Magnifier to view sample melting

MEL-TEMP®

FIGURE 5.29

(a) A solid sample is placed in a capillary tube for melting point determination. (b) A heated block apparatus heats the sample capillary tube, which can be viewed through a magnifier while monitoring the temperature via thermometer.

Courtesy of Gregory K. Friestad.

More sophisticated instrumental methods are usually not needed, but in some circumstances a technique called differential scanning calorimetry (DSC) can be used for a more rigorous quantitative analysis of phase transitions. In DSC, a sample along with a reference are heated slowly, and the amount of energy (heat) that each absorbs is measured over time. While the sample is undergoing a phase transition (i.e., at the melting point), it absorbs more energy than the reference. The energy input, or heat flow, is plotted versus temperature, and the melting point appears as a peak in the plot. DSC is used in the analysis of liquid crystals, polymers, and pharmaceuticals, and is also a tool for monitoring protein denaturation in biochemistry.

USING THE MELTING POINT TO ASSESS PURITY

Dissolved solutes (including impurities) lead to melting point depression. If the experimental melting point is more than 2–3°C less than the literature value for that compound, it may be considered impure. Usually, this melting point depression is also accompanied by a broadening of the melting range. Greater amounts of impurities lead to more melting point depression, perhaps by 20°C or more for a very impure sample, along with a more dramatic broadening of the range. Pure substances generally melt over a 1–2°C range. It is impractical to use melting point to quantify impurities in most situations, because there may be more than one impurity and their identities may be unknown. In some cases the impurity may not cause a problem for subsequent use of the compound, but if increased purity is needed, recrystallization is recommended.

USING THE MELTING POINT FOR IDENTIFICATION

To confirm the identity of a compound synthesized in the lab, or to identify an unknown compound, the experimental melting point range is simply compared with literature data. Impurities cause melting point depression, so for an impure compound, the experimental melting point may appear lower than the literature value. If the experimental melting point range is broader than two degrees, then melting point depression may also be occurring, and the melting point may be an unreliable means of identification. Melting point data are available for many thousands of organic compounds, and may be found in journal articles, handbooks, online databases, or other sources. Tables of melting point data for compounds (and their crystalline derivatives) are especially useful for identifying unknowns. These are usually organized by functional group, so the functional group must be known in order to use such tables.

MIXED MELTING POINT

If the identity of an unknown is narrowed to a couple of possibilities with similar melting points, and standards of high purity are available, then a mixed melting point determination can identify the compound. Each standard is separately mixed with the sample, and a melting point is determined on each mixture. The mixture that shows no melting point depression is the one in which the sample and standard are identical.

5.6C Density Measurement

Occasionally, the density of an organic liquid can be used to distinguish between alternative structures. For example, the boiling points of 1-bromo-4-chlorobenzene and 1-fluoro-4-iodobenzene are close enough that measuring them might be inconclusive (**Table 5.3**).

Their densities, though, are significantly different. Densities can be very quickly and conveniently measured, with accuracy to the second decimal place, simply by transferring liquid from a plastic 1.00-mL graduated syringe into a tared vial or small flask. Take care to read the start and end points on the syringe graduations to two significant figures (it doesn't have to be the full 1.00 mL, as long as the exact volume is known). Weigh the flask immediately to avoid loss to evaporation, and then calculate density by dividing the mass by the volume (g/mL). More precise measurements are possible using larger volumetric pipets, but higher precision is usually unnecessary for common organic lab operations.

TABLE 5.3

Properties of Two Halogenated Benzenes

COMPOUND	bp (°C)	DENSITY (g/mL)
1-Bromo-4-chlorobenzene	196	1.576
1-Fluoro-4-iodobenzene	189	1.925

5.7 POLARIMETRY

Most light is unpolarized, meaning that its photons have electric field oscillations in all possible directions. Polarized lenses restrict the light that can pass through the lens so that only the light oscillating in one plane can pass (**Figure 5.30**). The light that passes through is then called *plane-polarized light*. For a beam of light to pass through two such lenses, you would need to rotate one of the lenses until the angle of the polarization matches. This simple operation can be viewed using two lenses from a pair of polarizing sunglasses.

When plane-polarized light is passed through a sample of water or hexane (or any other achiral material), the plane on which it's polarized does not change. However, when plane-polarized light is passed through a solution of naturally occurring glucose (or any other chiral material), the plane-polarization is rotated

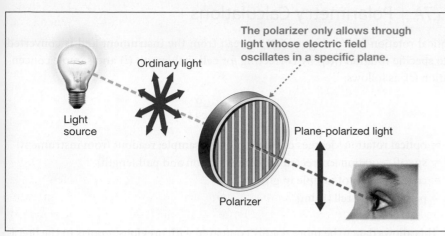

The polarizer only allows through light whose electric field oscillates in a specific plane.

Ordinary light

Light source

Plane-polarized light

Polarizer

FIGURE 5.30

A polarizing lens restricts light so that only the light oscillating in a certain plane can pass.

Karty, J. *Organic Chemistry: Principles and Mechanisms*, 3rd ed.; W. W. Norton: New York, 2022; p 245. Person's eyes: BLACKDAY/Shutterstock.

to a new angle (**Figure 5.31**). This is called **optical rotation**. The magnitude of that rotation can be measured on an instrument called a *polarimeter*. When the measurement is taken under standardized conditions of concentration, temperature, and wavelength, the optical rotation can be used to calculate **specific rotation**, a characteristic property of glucose.

This property is called *optical activity*, and a compound with optical activity is said to be *optically active*. Optical activity is closely associated with the stereochemistry of organic compounds. Two compounds that are related as enantiomers (nonsuperimposable mirror images) rotate plane-polarized light with equal magnitudes but in opposite directions. That is, one enantiomer rotates plane-polarized light clockwise (+), whereas the other enantiomer rotates it counterclockwise (−). When the specific rotation is reported in the literature, it is reported with both a sign and a magnitude. Thus, data from a laboratory sample can be correlated with literature data. Such a correlation can provide evidence of the identity of the compound, and enantiomeric purity can be determined using calculations described in the next section.

<< optical rotation
In polarimetry, the number of degrees by which a sample rotates plane-polarized light.

<< specific rotation
In polarimetry, an optical rotation value that has been corrected for wavelength, concentration, and path length, allowing comparison of optical rotation data despite differences in these quantities.

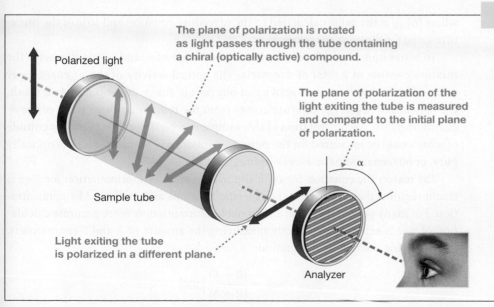

The plane of polarization is rotated as light passes through the tube containing a chiral (optically active) compound.

Polarized light

The plane of polarization of the light exiting the tube is measured and compared to the initial plane of polarization.

α

Sample tube

Light exiting the tube is polarized in a different plane.

Analyzer

FIGURE 5.31

Plane-polarized light passes through an optically active sample solution, and a polarizing lens is rotated until it is aligned to allow light through.

Karty, J. *Organic Chemistry: Principles and Mechanisms*, 3rd ed.; W. W. Norton: New York, 2022; p 246. Person's eyes: BLACKDAY/Shutterstock.

5.7A | Polarimetry Calculations

Optical rotation (α; see Figure 5.31) is read from the instrument and is converted into specific rotation ($[\alpha]$) by correcting for cell pathlength (l) and sample concentration (c), as follows:

$$[\alpha] = \frac{\alpha}{c \cdot l}$$

α = optical rotation (degrees of rotation of the sample; readout from instrument)
$[\alpha]$ = specific rotation (corrected for concentration and pathlength)
c = concentration of sample in g/mL
l = pathlength of cell (1 dm)

This allows data to be more readily compared with previous reports in the literature. Specific rotation also depends on temperature and the wavelength of light (the "sodium D-line" at 589 nm is typically used). In the literature, data are reported as $[\alpha]_D^{25}$, where the subscript is the wavelength (here the sodium D-line), and the superscripted number is the temperature (in this case, 25°C).

Measuring the optical rotation can be used to determine the enantiomeric ratio of a sample that is a mixture of enantiomers. A racemic mixture (equal parts of both enantiomers) is optically inactive, because the two opposing optical rotations cancel each other out. If there is an excess of one enantiomer, then the sample is optically active. A single enantiomer of 100% enantiomeric purity will have a specific rotation that matches the literature value for the enantiomer, and the sign of that rotation can be used to determine which enantiomer it is.

When a mixture is nonracemic, however, the *enantiomeric excess* (%ee) describes its enantiomeric purity, which can be approximated by comparing the observed calculated specific rotation with that found in the literature for the known pure enantiomer,

$$\%ee \approx \% \text{ optical purity} = ([\alpha]_{calc}/[\alpha]_{lit}) \times 100$$

where $[\alpha]_{calc}$ is the value calculated in the preceding equation and $[\alpha]_{lit}$ is the literature value for the pure enantiomer.

To better understand %ee, imagine a 4:1 mixture of + and − enantiomers. The mixture consists of a total of five parts. The optical activity of the − enantiomer is canceled out by the optical activity of one part of the + enantiomer. As a result, the optical activity of the mixture comes from the remaining three parts of the + enantiomer. The specific rotation of this sample will be 3/5 or 60% of the magnitude of what would be measured for the pure + enantiomer. This sample is 60% optically pure, or 60% enantiomeric excess (abbreviated %ee).

The reason the equation for optical purity is only an approximation for %ee is that it requires the assumption that specific rotations are unaffected by concentration. For many purposes, that is a reasonable assumption. A more accurate calculation of %ee is achieved by directly measuring the amount of R and S enantiomers, then applying the following equation:

$$\%ee = \frac{(R - S)}{(R + S)}(100)$$

The accurate measurement of R and S enantiomer ratios is generally achieved by use of chiral chromatography with a stationary phase that has chiral structure. This can be achieved with gas chromatography (GC) or high-performance liquid chromatography (HPLC). If the stationary phase is chiral, then the two enantiomers will associate with the stationary phase with different affinities, in which case they will pass through the chromatographic column at different rates. A detector can integrate the peaks corresponding to the two enantiomers and determine their ratio.

5.7B │ Sample Preparation for Polarimetry

 Polarimetry

Polarimetry is usually measured on dilute solutions—namely, concentrations near 1 mg/mL. Typical organic solvents for polarimetry are methanol (CH_3OH) and chloroform ($CHCl_3$), which should be handled in the fume hood, although very polar substances such as sugars and amino acids are frequently measured in aqueous solution. If you want to compare the optical rotation value you obtain experimentally with a literature value, choose your solvent and concentration to match the literature report, insofar as possible.

The concentration is important, so it must be accurately determined, usually to three significant figures. This means that both mass and volume must be measured to three significant figures. The best way to do this is to transfer the sample to a tared volumetric flask, measuring the mass of sample on an analytical balance. The volumetric flask or tube should be chosen so that its volume is slightly larger than the volume of the polarimetry cell. Dilute the sample with the chosen solvent, mix thoroughly, and fill the volumetric flask up to the marked volume line. Invert the stoppered flask several times to ensure that the solution is homogeneous. Alternatively, the mass of the sample may be determined in a vial or flask, then transferred quantitatively to the volumetric flask using at least three small portions of solvent.

Clean the polarimetry cell carefully before and after use, and be sure that any cleaning solvents are completely removed from the cell. When transferring the sample from the volumetric flask to the polarimetry cell, no further solvent can be used, or the concentration data will become inaccurate. If there is sufficient sample available, a small portion of the sample can be used to rinse the polarimetry cell before filling it. Before measuring optical rotation, inspect the cell visually to make sure the light path is not obstructed by bubbles.

Polarimetry is a nondestructive measurement, so a small precious sample may be recovered, if necessary, and used for other analytical or synthetic purposes.

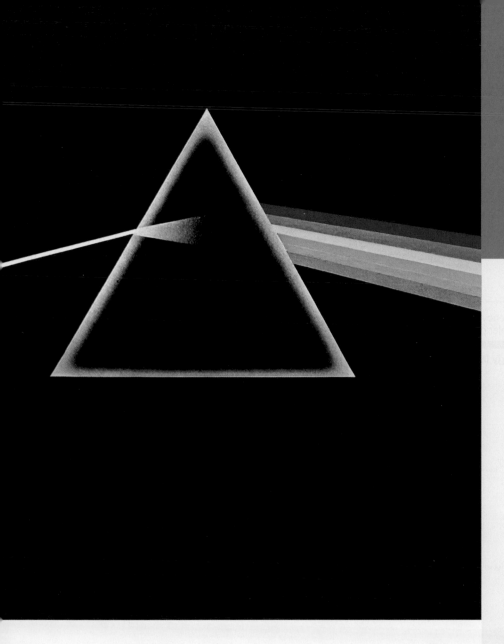

6

LEARNING OBJECTIVES

- Apply mathematical relationships that connect wavelength, frequency, energy, absorbance, and concentration.

- Relate infrared and ultraviolet radiation absorbance to structural features of organic compounds.

- Differentiate and identify organic compounds, using IR and UV data.

- Report IR and UV data in the appropriate formats.

Absorption Spectroscopy in Organic Chemistry

DIFFERENT COLORS, DIFFERENT ENERGIES

Visible white light contains different colors of light at different wavelengths, as part of a broader electromagnetic spectrum.

Records/Alamy Stock Photo.

INTRODUCTION: THEORY OF ABSORPTION SPECTROSCOPY

Absorption spectroscopy is a valuable tool for structure determination in organic chemistry. Absorptions of radiation from various parts of the electromagnetic spectrum are associated with certain structural characteristics in organic compounds (**Figure 6.1**). For example, ultraviolet (UV) and infrared (IR) light have different frequencies, so irradiating organic compounds with these forms of light impacts the structure in different ways. The absorption of ultraviolet/visible (UV/vis) light promotes an electron from one molecular orbital to a higher-energy orbital, whereas the absorption of infrared light causes changes in **molecular bond vibrations** such as stretching and bending motions.

molecular bond vibrations >>
Changes in the lengths and angles of bonds, stimulated by absorbance of light in the infrared region of the electromagnetic spectrum.

	Higher frequency (ν) Shorter wavelength (λ) Higher energy (E)				Lower frequency (ν) Longer wavelength (λ) Lower energy (E)
ν (cm^{-1})	10^6	10^4	10^2	10^0	10^{-1}
λ	10 nm	200–700 nm	1–100 μm	10 mm	100 mm
Name	X-rays	Ultraviolet–visible	Infrared	Microwave	Radiofrequency (RF)
Molecular transition process	Excitation of core atomic electrons	Excitation of molecular valence electrons	Stretching and bending of molecular bonds	Rotational transitions of bonds	Transitions of nuclear spin magnetic dipoles
ΔE (kJ mol^{-1})	1200	300	10	10^{-4}	10^{-6}
		UV–vis spectroscopy	**IR spectroscopy**		**NMR spectroscopy**

FIGURE 6.1

The electromagnetic spectrum and its relationship to the energies associated with molecular transition processes.

These structural features respond to light from different parts of the electromagnetic spectrum. To understand why, we first review the basics of the wave and energy properties of light. Frequency (ν = the number of waves that pass a certain point during a given time period) is related to wavelength (λ = the distance between one point on a wave to the same point on an adjacent wave) by the equation,

$$\nu = c/\lambda$$

where c = speed of light = 3×10^8 m s^{-1}. Frequency is related to energy E by the equation,

$$E = h\nu$$

where h = Planck's constant = 6.6×10^{-34} J·s. Combining these two equations, we get

$$E = hc/\lambda$$

Therefore, the energy associated with light is *directly* proportional to its *frequency*, and *inversely* proportional to its *wavelength*. In other words, higher frequencies (shorter wavelengths) of light have higher energies, whereas lower frequencies (longer wavelengths) have lower energies.

According to the electromagnetic spectrum in Figure 6.1, UV light has a shorter wavelength than IR light. Promoting an electron to an unfilled molecular orbital is matched with the amount of energy imparted by one photon of UV light, so that is the transition that happens when the photon is absorbed. Meanwhile, the energy associated with vibration of specific bonds is matched with specific amounts of energy of photons in the IR region of the electromagnetic spectrum. When an IR photon is absorbed, vibrations are excited to a higher-energy state; in other words, the energy of the photon is converted to vibrational energy within the molecular structure. Specific energies in the IR region correspond to various stretching and bending vibrations of bonds within an organic compound. Excitations of bond rotations, on the other hand, require less energy than bond stretching and bending; these are associated with absorption of energy in the microwave region of the spectrum.

Lower still on the energy scale of the electromagnetic spectrum is radiofrequency radiation. This is the frequency range you use when you listen to your local FM radio stations (about 100 MHz). Energy associated with this radiation is matched with transitions of magnetic dipoles associated with nuclear spins, and is used for nuclear magnetic resonance (NMR) spectroscopy, which we will discuss further in Chapter 7.

In the sections that follow, we explain how to use molecular transitions associated with ultraviolet, visible, and infrared light for the analysis and characterization of samples prepared in the organic laboratory. Specifically, this chapter describes how to prepare samples, acquire spectra, interpret results, and report the data in a scientific report.

6.1 ULTRAVIOLET–VISIBLE SPECTROSCOPY

The absorption of light in the ultraviolet (UV) and visible (vis) region ($\lambda = 200$–700 nm) is associated with an **electronic transition** from an occupied lower-energy molecular orbital to an unoccupied higher-energy one in a molecule. When light's specific wavelength and energy match the difference in energy between these two orbitals, a photon can be absorbed by the molecule and an electron is promoted to the higher-energy orbital.

Instruments that detect the absorbance of UV–vis light work by passing the light through a cell (called a cuvette) containing a solution of the sample in a solvent that is transparent in the 200–700 nm range. Typical solvents include water, methanol, hexane, and acetonitrile. The sample is prepared by dissolving a known mass of the sample in such a solvent, then diluting it to a known volume, to achieve a concentration of about 10^{-4} M or lower. To achieve precision at very low concentrations, volumetric flasks and volumetric pipets may be used, with quantitative **serial dilutions** as needed. Once prepared, the sample is placed in a cuvette, and the spectrum is acquired. Interference from the solvent may be removed by acquiring a reference spectrum of a blank cuvette containing only the solvent, either prior to acquiring the sample spectrum (within a single beam instrument) or simultaneously (with a double beam instrument). The spectrum of the blank solvent is electronically subtracted from the sample spectrum.

The UV–vis absorbance can be detected at a single chosen wavelength, or by using a photodiode array detector, at all wavelengths in the region simultaneously. The latter results in a spectrum plotted with wavelength on the x axis and absorbance on the y axis (**Figure 6.2**). Notice that the absorbance is broad with respect to the wavelength. When data are presented in tables or databases, the wavelength is reported at its maximum absorbance (λ_{max}).

<< **electronic transition**
Movement of an electron from one orbital to another one of higher energy, stimulated by the absorbance of light in the ultraviolet–visible region of the electromagnetic spectrum.

<< **serial dilutions**
Using a portion of a diluted sample as the concentrate in a subsequent dilution, multiplying the dilution effect.

FIGURE 6.2

UV spectrum of 1,3-butadiene.

Karty, J. *Organic Chemistry: Principles and Mechanisms*, 3rd ed.; W. W. Norton: New York, 2022; p 802.

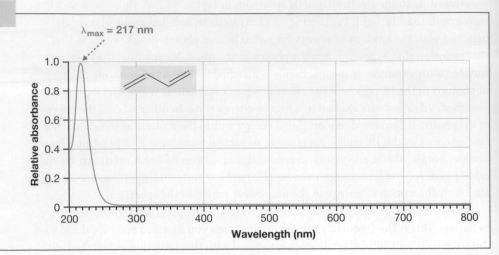

6.1A Correlating λ_{max} with Structural Features

Organic compounds that have useful UV–vis spectra in the 200–700 nm region generally have functional groups with a π bond. This is because π bonding and antibonding orbitals are close enough in energy to allow light in this range to be absorbed. Exciting an electron in ethylene from π to π^* orbitals corresponds to absorption at $\lambda_{max} = 171$ nm, just outside the useful range. Conjugation of the π bond, though, causes this energy gap ΔE to diminish (**Figure 6.3**), so 1,3-butadiene absorbs at $\lambda_{max} = 217$ nm. Each successive π bond in conjugation further diminishes the gap between π and π^*, thereby moving the absorbance to longer wavelengths.

FIGURE 6.3

The π molecular orbitals of (a) ethylene, (b) butadiene, and (c) 1,3,5-hexatriene, showing how conjugation decreases the energy gap between π (HOMO) and π^* (LUMO) orbitals.

Karty, J. *Organic Chemistry: Principles and Mechanisms*, 3rd ed.; W. W. Norton: New York, 2022; p 808.

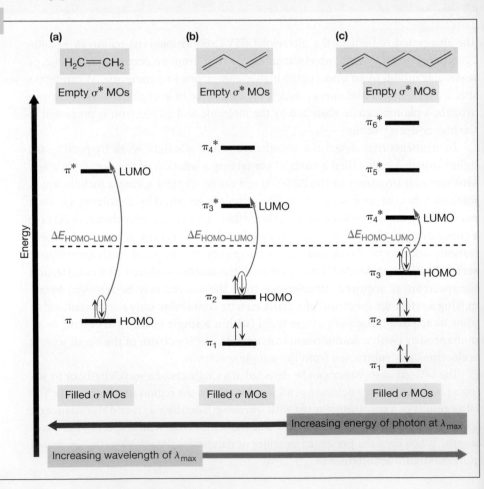

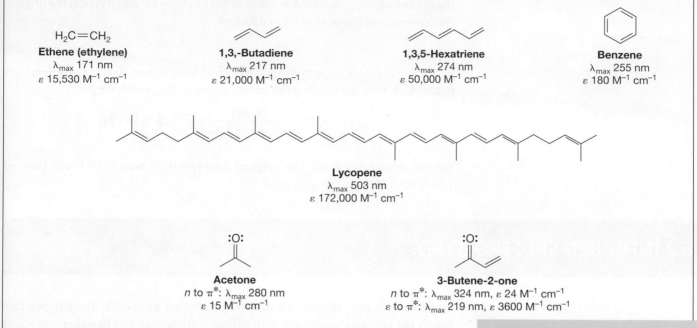

FIGURE 6.4

Selected UV–vis data for a variety of organic compounds.

For example, *trans*-1,3,5-hexatriene absorbs at $\lambda_{max} = 274$ nm and lycopene at $\lambda_{max} = 503$ nm (**Figure 6.4**). Lycopene is found in ripe tomatoes and contributes to their intensely red color. The π bonds at the ends of lycopene are not conjugated with the rest, so they do not affect its λ_{max}.

Lone pairs of electrons associated with an atom involved in a π bond populate non-bonding orbitals, or *n* orbitals, and these can also be promoted to higher-energy **π^* orbitals** by absorption in the UV–vis region. This is referred to as an *n* to π^* transition. As with alkenes, conjugation also causes these transitions to occur with longer wavelengths. This can be seen by comparing the λ_{max} for acetone and 3-butene-2-one in Figure 6.4.

<< **π^* orbital**
Antibonding orbital generated from combination of the *p* orbitals of two atoms.

6.1B Using UV–Vis Spectroscopy for Quantitative Analysis

Another feature of the UV–vis spectrum is the intensity of each peak, or its absorbance (*A*), measured on the *y* axis (see Figure 6.2). Absorbance is related to the **molar absorptivity or molar extinction coefficient** (ε), of the compound, which is a constant that is characteristic of the specific compound. A compound that absorbs in the visible light region with a high ε, such as lycopene, will appear strongly colored even at lower concentrations. Absorbance is also dependent on the pathlength of the sample cell or cuvette, in centimeters (*b*), and the concentration of the sample (*c*). These three terms relate to absorbance according to Beer's law:

$$A = \varepsilon \cdot b \cdot c$$

When acquiring data for known compounds, ε and *b* are usually known, and *A* is measured by the instrument, leaving *c* as the only variable. Thus, Beer's law can be used to calculate *c*, the concentration of an analyte in solution, from a UV–vis spectrum.

<< **molar absorptivity or molar extinction coefficient**
A quantity that describes how strongly a compound absorbs light; it varies among compounds and is a characteristic physical property of compounds that absorb light. It appears as ε, a constant in Beer's law $A = \varepsilon \cdot b \cdot c$. This constant enables measurements of absorbance (*A*) to be used to calculate sample concentration (*c*).

Worked Example

The extraction of lycopene from vegetable material yielded a solution of lycopene. After diluting to 1/1000th of the original concentration, UV–vis spectroscopy of the diluted solution showed $A = 0.200$ at 503 nm ($b = 1.0$ cm). What is the original concentration of lycopene prior to dilution?

Solution

According to Figure 6.4, $\varepsilon = 172{,}000$ M^{-1} cm^{-1} for lycopene. Rearranging Beer's law to $c = A/\varepsilon b$, then substituting the known values of A, ε, and b, we get

$$c = \frac{A}{\varepsilon b} = \frac{0.200}{\left(172{,}000 \text{ M}^{-1} \text{cm}^{-1}\right)} = 1.16 \times 10^{-6} \text{ M}$$

for the diluted solution. The original concentration was 1000 times that, or 1.16×10^{-3} M.

6.2 INFRARED SPECTROSCOPY

Functional groups absorb infrared (IR) radiation at specific frequencies that match the energies associated with different stretching and bending vibrations of their bonds (**Figure 6.5**). The wavelength range (micrometers) is slightly longer than that of visible light. Expressed in wavenumbers (ν) with units of waves per centimeter (cm^{-1}), the frequency range of infrared spectroscopy is approximately 500–4000 cm^{-1}. Absorbances at higher wavenumbers are connected to vibrations of higher energy, and this in turn correlates to structural features of specific bonds. Various types of functional groups have vibrations at frequencies that are broadly distributed across this range. Infrared spectroscopy is therefore an excellent method for identifying organic functional groups in experimental samples.

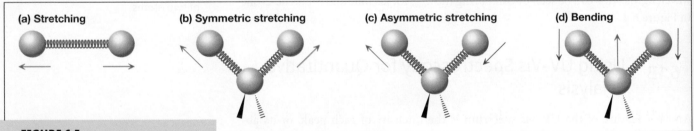

FIGURE 6.5

Various vibrations of organic compounds.

Gilbert, T.; Kirss, R.; Bretz, S.; Foster, N. *Chemistry*, 6th ed.; W. W. Norton: New York, 2020; p 409.

6.2A Infrared Absorption and Molecular Transitions

The bonds of organic molecules stretch and bend at specific vibrational frequencies, and the physics of this can be related to that of springs with weights attached. Hooke's law describes the vibration of a spring in terms of the masses of the weights and the stiffness of the spring (**Figure 6.6**). The frequency of vibrations is proportional to the stiffness of the spring, and inversely proportional to the masses of the weights. Hooke's law can be applied to covalent bonds as shown in Figure 6.6.

Hooke's law predicts, all other things being equal, that bonds between lighter atoms will vibrate at higher frequency than those between heavy atoms. And, all other things being equal, stronger bonds will have higher frequency vibrations than weaker bonds. These predictions are confirmed by measurements of the stretching frequencies of a C—H bond (3000 cm⁻¹) versus a C—D bond (2200 cm⁻¹), and the stretching frequencies of a C=C bond (1650 cm⁻¹) and a C≡C bond (2200 cm⁻¹). A wide range of organic functional groups and bonds are correlated to various vibrational frequency ranges, as observed in IR spectra (**Figure 6.7**). Tables of such data can be useful for specifying the functional group and its neighboring structural features in more detail (**Table 6.1**). Intensities of the peaks are noted as strong (s), medium (m), or weak (w) in the table. These intensities are related to bond polarization; for example, peaks related to C=O bond vibrations have strong intensities while C=C bonds are weaker.

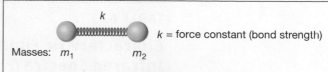

k = force constant (bond strength)

Masses: m_1 m_2

Vibrational frequency v is given by Hooke's law:

$$v = \frac{1}{2\pi c}\left(\frac{k}{\mu}\right)^{1/2}$$

where c = speed of light, k = force constant, μ = reduced mass =

$$\frac{m_1 m_2}{m_1 + m_2}$$

FIGURE 6.6

Relationships of atomic masses and bond strengths to vibrational frequency.

FIGURE 6.7

Representative IR spectrum showing the parts of the spectrum where various types of bond vibrations absorb in the infrared frequency range. Q = C, N, or O.

Karty, J. *Organic Chemistry: Principles and Mechanisms*, 3rd ed.; W. W. Norton: New York, 2022; p 777.

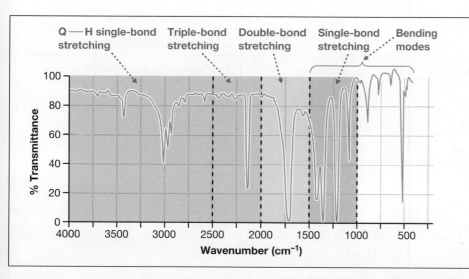

TABLE 6.1

Characteristic Functional Group Absorbances in Infrared Spectra

FREQUENCY (cm⁻¹)	BOND	FUNCTIONAL GROUP
3640–3610 (s, sh)	O—H stretch, free hydroxyl	Alcohols, phenols
3500–3200 (s, br)	O—H stretch, H–bonded	Alcohols, phenols
3400–3250 (m)	N—H stretch	Primary, secondary amines, amides
3300–2500 (m, br)	O—H stretch	Carboxylic acids
3330–3270 (m, s)	—C≡C—H: C—H stretch	Alkynes (terminal)
3100–3000 (m)	C—H stretch	Aromatics
3100–3000 (m)	=C—H stretch	Alkenes
3000–2850 (m)	C—H stretch	Alkanes

(continued)

TABLE 6.1

Characteristic Functional Group Absorbances in Infrared Spectra (continued)

FREQUENCY (cm⁻¹)	BOND	FUNCTIONAL GROUP
2830–2695 (m)	H—C=O: C—H stretch	Aldehydes
2260–2210 (v)	C≡N stretch	Nitriles
2260–2100 (w)	—C≡C— stretch	Alkynes, non-symmetrical
1760–1665 (s)	C=O stretch	Carbonyls (general)ᵃ
1760–1690 (s)	C=O stretch	Carboxylic acids
1750–1735 (s)	C=O stretch	Esters, saturated aliphatic
1740–1720 (s)	C=O stretch	Aldehydes, saturated aliphatic
1730–1715 (s)	C=O stretch	α,β-Unsaturated esters
1715 (s)	C=O stretch	Ketones, saturated aliphatic
1710–1665 (s)	C=O stretch	α,β-Unsaturated aldehydes, ketones
1680–1630 (s)	C=O stretch	Amides
1680–1640 (m)	—C=C— stretch	Alkenes
1600–1585 (m)	C—C stretch (in–ring)	Aromatics
1550–1475 (s)	N—O asymmetric stretch	Nitro compounds
1500–1400 (m)	C—C stretch (in–ring)	Aromatics
1470–1450 (m)	C—H bend	Alkanes
1370–1350 (m)	C—H rock	Alkanes
1360–1290 (m)	N—O symmetric stretch	Nitro compounds
1335–1250 (s)	C—N stretch	Aromatic amines
1320–1000 (s)	C—O stretch	Alcohols, carboxylic acids, esters, ethers
1300–1150 (m)	C—H wag (—CH₂X)	Alkyl halides
1250–1020 (m)	C—N stretch	Aliphatic amines
1000–650 (s)	=C—H bend	Alkenes
950–910 (m)	O—H bend	Carboxylic acids
910–665 (s, br)	N—H wag	Primary, secondary amines
900–675 (s)	C—H "oop"	Aromatics
850–550 (m)	C—Cl stretch	Alkyl halides
725–720 (m)	C—H rock	Alkanes
700–610 (br, s)	—C≡C—H: C—H bend	Alkynes
690–515 (m)	C—Br stretch	Alkyl halides

(s) = strong, (m) = medium, (w) = weak, (br) = broad, (sh) = sharp, (v) = variable
ᵃConjugation of other unsaturated groups with the carbonyl will lower the C=O frequency by 20–40 cm⁻¹.

6.2B Acquiring an Infrared Spectrum

The spectrum is obtained by passing IR radiation through a sample of the compound and measuring the light that is transmitted through the sample. The result is plotted in percent **transmittance** (%T) versus frequency (**Figure 6.8**), with the frequency expressed in units of wavenumbers (waves per centimeter, cm^{-1}). It is useful to understand how transmittance relates to absorbance. Transmittance and percent transmittance are calculated as follows:

$$T = I/I_0$$
$$\%T = (I/I_0)100$$

where I_0 is the initial light intensity and I is the intensity of light after it has passed through the sample. Transmittance is inversely related to absorbance through a logarithmic function:

$$A = \log_{10}(1/T)$$
$$A = 2 - \log_{10}(1/\%T)$$
$$\%T = (1/10^A)100$$

From these equations, we can see that when transmittance is 100%, the absorbance is 0, and working in the other direction, an absorbance of 2 corresponds to 1% transmittance.

<< transmittance
A fraction I/I_0 of light intensity remaining after it has passed through a sample (I), relative to its initial intensity (I_0). This fraction, expressed as a percentage (%T), is the unit commonly found on the y axis of an infrared spectrum.

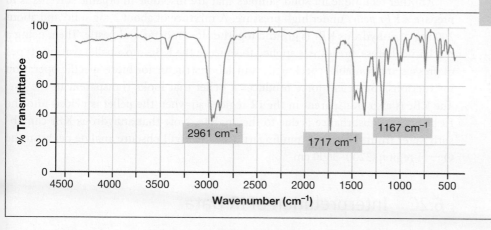

FIGURE 6.8

A typical infrared spectrum.

Karty, J. *Organic Chemistry: Principles and Mechanisms*, 3rd ed.; W. W. Norton: New York, 2022; p 774.

Usually it is desirable to subtract background peaks that mainly result from humidity and/or CO_2 in the air. A background spectrum is recorded and the instrument can subtract it from the sample spectrum to obtain a cleaner result. Using the instrument software, the peaks in the resulting spectrum can be labeled in units of cm^{-1}, and a spectrum can be printed so that it can be attached to a lab report.

The best quality spectrum is one in which the largest peak is in the range of 10–50%T. If the largest peak is above 50%T, then it is likely that smaller peaks may be lost in instrumental noise. If there are peaks at 0% transmittance, then there is too much sample and the spectrum gives little indication of the relative intensity of the peaks. In either case, adjust the amount of sample and acquire another spectrum.

SAMPLE PREPARATION FOR LIQUIDS

The simplest liquid sample preparation procedure is to place a drop of a liquid sample between two salt (NaCl) plates, forming a thin film of the sample, and then place the sandwiched sample in the path of the IR beam. If there is too much sample (peaks bottom out at 0%T), take the plates apart, wipe one off with a tissue to remove some sample, and acquire the spectrum again. To avoid scratching, do not contact the plate with a pipet. *CAUTION: Water will dissolve salt plates or damage them. Avoid exposing salt plates to water, whether it is present in the sample itself, or moisture from your fingers. Avoid contact with the faces of the salt plates.*

SAMPLE PREPARATION FOR SOLIDS

For solids, the same procedure may be used by placing a couple of milligrams of solid (the tip of a spatula) on the salt plate, adding a drop of $CHCl_3$ to make a solution, then sandwiching the solution between two salt plates. For solids that are sparingly soluble in $CHCl_3$, a spectrum may sometimes be obtained by heating the solid briefly with $CHCl_3$ in a test tube and placing a drop of the supernatant solution on the salt plate. After the $CHCl_3$ evaporates, the sample remains as a thin film. If this is still unsuccessful, a nujol mull can be prepared by grinding the solid into a fine powder and mixing it with nujol. Nujol is a mineral oil, with strong absorbances in the C—H region that can interfere with the spectrum. If $CHCl_3$ is used, a background spectrum should be obtained with just $CHCl_3$ so that it can be subtracted from the spectrum.

A procedure called *attenuated total reflectance (ATR)* provides another option that is particularly useful for insoluble solids. In ATR, which requires a special attachment for the instrument, the solid is pressed against a small window made of a germanium or zinc selenide crystal. The IR beam passes through that window and bounces off the surface of the solid. Small amounts of IR radiation are absorbed at the surface of the solid sample, resulting in an infrared spectrum from solids that otherwise would be difficult to analyze. The ATR method also works well for liquid samples.

Another technique for solid samples that are insoluble in organic solvents is to prepare a *KBr pellet* under high pressure. A mixture of about 2 mg solid and about 200 mg of anhydrous KBr is ground together in a mortar and pestle. Then, using a mechanical press, pressure is applied to squeeze the KBr into a pellet (various types of presses are available; check with your lab instructor for more specific operation instructions). Sufficient force produces a pellet that looks transparent to the eye. The KBr is also transparent in the IR region, so when the pellet is placed into the IR beam, any absorbance is due to the analyte. Note that anhydrous KBr can absorb water from the air or from fingertips, which will obscure analyte peaks in the O—H region ($3200–3600$ cm^{-1}).

6.2C | Interpretation of IR Data

Within the IR spectrum, there are two main ranges that are used for different purposes. The range from $1650–4000$ cm^{-1} provides the most useful diagnostic information for identifying functional groups present in the sample. The range from $500–1650$ cm^{-1} is often called the **fingerprint region**, and is only useful in certain circumstances. The fingerprint range often contains too many peaks to interpret each one individually for an unknown sample. Instead it is most useful when comparing the IR spectrum of a sample of high purity with a known literature spectrum to confirm that they match. As its name suggests, it is rare to have two compounds with an IR spectrum showing the same "fingerprint" of peaks in the fingerprint region.

CHARACTERIZATION OF A KNOWN COMPOUND BY IR

To characterize the features of a known compound using IR, perform the following steps.

1. Know the structure of the compound you are attempting to characterize.

2. Examine the structure and identify the functional groups you would expect to see in the IR.

fingerprint region >>
The lower-energy portion, or smaller wavenumber portion, of an infrared spectrum, typically 700–1600 cm^{-1}. This region of the spectrum contains many peaks that are sometimes difficult to assign, yet are useful in identification of an unknown by direct comparison of its fingerprint region to that of a known standard.

3. Determine the position (cm^{-1}) at which you would expect to find each peak for a given functional group. For some functional groups (e.g., carboxylic acid, nitro, etc.), you should expect to see several diagnostic peaks. If the compound is a reaction product, focus most attention on those peaks that correspond to the functional group that changed during the reaction; these peaks will be diagnostic for the success or failure of the reaction.

4. Obtain a good clear IR of your sample. If the compound is a solid, attenuated total reflectance (ATR) can be used. If ATR is not available, the spectrum can be obtained using a thin film of a solution of the solid in CHCl$_3$, placed between salt plates. A spectrum may also be obtained from a "nujol mull"—a slurry of the solid in mineral oil. The largest peak should have a nonzero value for % transmittance (%T).

5. Considering your preliminary expectations, identify those peaks in the spectrum that are diagnostic for your compound.

6. If unexpected peaks appear in your IR spectrum, you will need to explain them. Common impurities include unreacted starting materials, reagents, water, and solvents.

7. The IR spectra for millions of compounds can be found in the primary chemical literature (journal articles) or in online databases, either in graphic form or as a listing of peak frequencies. These data may be compared with your spectrum in order to strengthen your confidence in structure assignments and the identification of impurities. For best comparison, choose a literature spectrum obtained using the same method (e.g., ATR, KBr pellet, solution, film, etc.).

8. If you know that your sample was a single compound of high purity, you may compare the fingerprint region of your IR spectrum with one from the literature to see if they match. For an impure compound there can be a lot of extraneous peaks in the fingerprint region, so use caution in interpreting this part of the spectrum.

EVALUATION OF AN UNKNOWN SAMPLE BY IR

Preliminary evaluation: strongly diagnostic peaks for functional groups.

1. Determine what you would expect to see (peak position, number of peaks) for each of the possible major organic functional groups in your sample. Consider also any special circumstances that could affect what you see (e.g., hydrogen bonding, amine substitution, etc.).

2. Obtain a good clear IR of your sample. If the compound is a solid, attenuated total reflectance (ATR) can be used. If ATR is not available, the spectrum can be obtained using a thin film of a solution of the solid in CHCl$_3$, placed between salt plates. A spectrum may also be obtained from a "nujol mull"—a slurry of the solid in mineral oil. The largest peak should have a nonzero value for % transmittance (%T).

3. Based on your IR spectrum, determine which functional groups may be present in your sample. You will probably be unable to narrow it down to a single group, but you should be able to eliminate a number of options based on peaks that are *not* present (e.g., if no C=O stretch is present, then esters, carboxylic acids, aldehydes, and ketones can be eliminated from consideration). Often, peaks in the range 3100–3600 cm^{-1} have characteristic shapes that are useful: U-shape for alcohol O—H, W-shape for NH$_2$, V-shape for NH, and a sharp spike for alkyne C—H.

Subsequent evaluation: peaks with more ambiguous interpretations.

1. Using the information you have obtained from functional group tests and other sources, reinspect your IR spectrum to gain additional support for your findings. Consider additional functional groups, including double and triple bonds, halides, nitro groups, ethers, etc. IR spectra are usually very complex. Do not try to read more into the spectrum than is actually there.

2. If the number of possible compounds is small, locate IR spectra in the chemical literature for each of the possibilities and compare all the peaks, including the fingerprint region. If your unknown sample is pure, its spectrum should closely match the known spectrum found in a book or online database. If there are more than a handful of possible compounds, it can be tedious to find a fingerprint match. First try to narrow the possibilities using other data, such as physical properties (mp, bp) or number of carbons.

6.2D Reporting IR Data

Report all information from your IR spectrum that is relevant to the identification and purity of the compound(s) that you have prepared. IR data should be reported in table format as shown in the example that follows. Peak positions should be reported in cm^{-1} with an assignment to a type of bond (e.g., O—H, C=O, etc.) and a specific functional group (ester, alcohol, etc.). When the absence of a peak is significant, mention in the report that this peak is absent and interpret what that absence implies.

Worked Example: IR Spectra of Ethyl Levulinate
When a student esterified levulinic acid (a carboxylic acid), the final product obtained was ethyl levulinate (**Figure 6.9**).

FIGURE 6.9

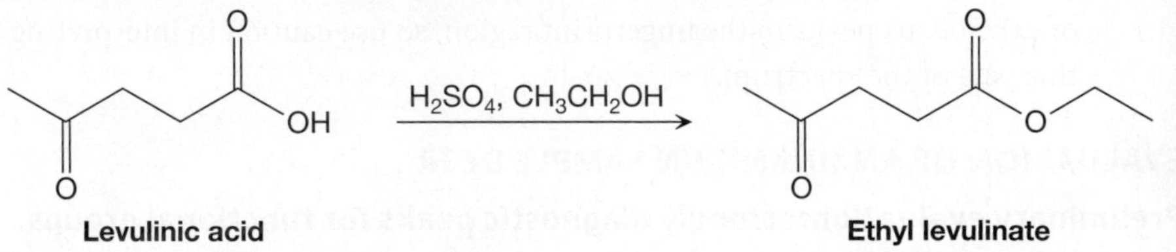

H_2SO_4, CH_3CH_2OH

Levulinic acid   **Ethyl levulinate**

The IR spectrum in **Figure 6.10** was obtained of the reaction product.

FIGURE 6.10

Spectral Database for Organic Compounds, SDBSWeb. National Institute of Advanced Industrial Science and Technology. https://sdbs.db.aist.go.jp/ (accessed July 2022). Reprinted by permission.

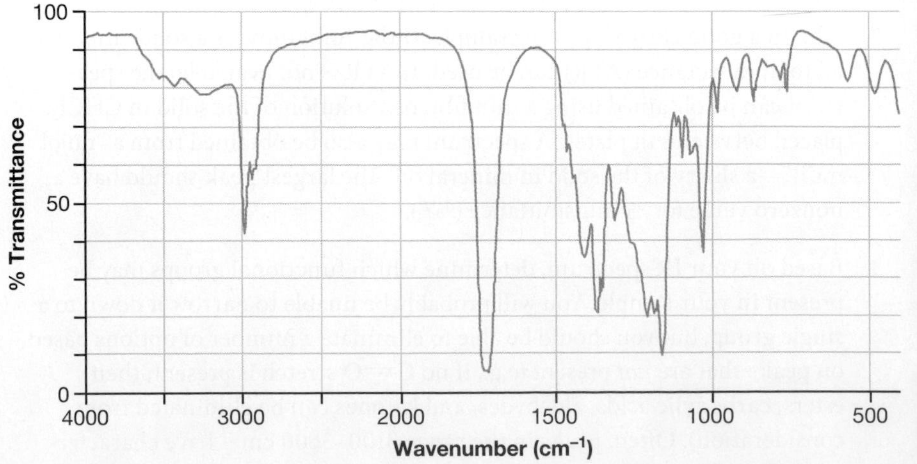

The student interpreted the spectrum and reported the data in the Results and Discussion section of the lab report, using data from **Table 6.2**.

TABLE 6.2

IR Data for Ethyl Levulinate

FREQUENCY (cm^{-1})	BOND	FUNCTIONAL GROUP
2950	C—H	Alkane
1740	C=O	Ester
1720	C=O	Ketone
1160	C—O	Ester

Note that the two carbonyl peaks appear to partially overlap. In the body of the report, the student discussed what was meant by the presence of each of the peaks indicated in the table, and also discussed the absence of a broad O—H peak from 3300–2500 cm^{-1}, which would be expected from the CO$_2$H functional group in the starting material. The student noted that the absence of this O—H absorbance indicated that the reaction was successful and had gone to completion. The small O—H absorbances in the range of 3200–3500 cm^{-1} were attributed to small amounts of water or ethanol contaminating the sample.

6.3 SAMPLE PROBLEMS

1. (a) Match the compounds in **Figure 6.11** with the spectra in **Figure 6.12**.
    (b) For each of the spectra in Figure 6.12, identify at least two peaks that are diagnostic for the structure you matched to it. Write the wavenumber of the peak and assign it to a specific bond vibration (e.g., C—H stretch).

FIGURE 6.11

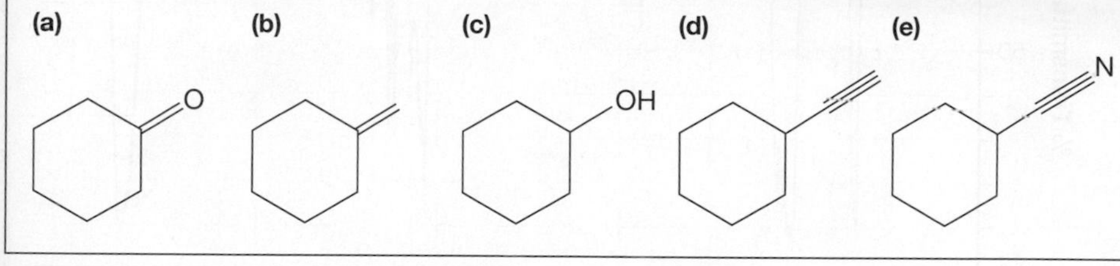

FIGURE 6.12

Spectral Database for Organic Compounds, SDBSWeb. National Institute of Advanced Industrial Science and Technology. https://sdbs.db.aist.go.jp/ (accessed July 2022). Reprinted by permission.

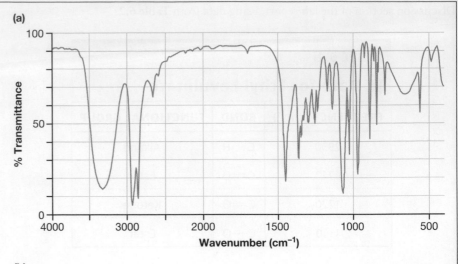

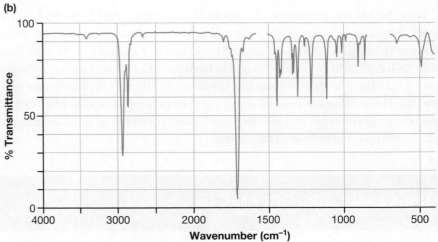

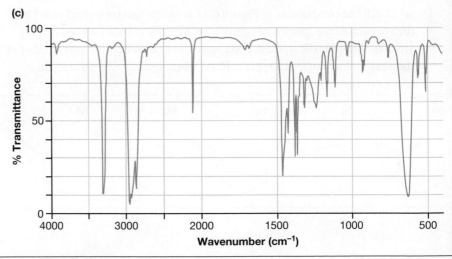

(continued)

(d)

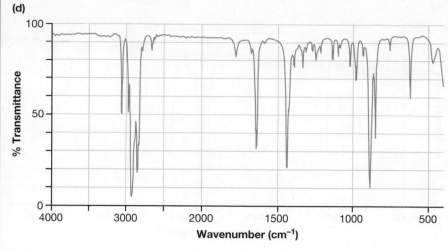

(e)

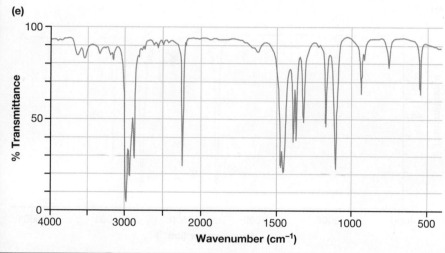

2. Propose a structure for each of the spectra in **Figure 6.13**, using the formula provided.

    a. C_7H_8O

    b. C_4H_9NO

    c. $C_4H_8O_2$

    d. C_7H_6O

FIGURE 6.13

When acquiring the IR spectrum of this compound under other conditions, the peak in the 3200–3600 cm^{-1} region appeared more broad and U-shaped.

Spectral Database for Organic Compounds, SDBSWeb. National Institute of Advanced Industrial Science and Technology. https://sdbs.db.aist.go.jp/ (accessed July 2022). Reprinted by permission.

(a)

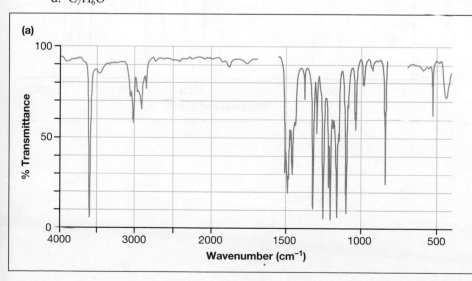

(continued)

FIGURE 6.13 *(continued)*

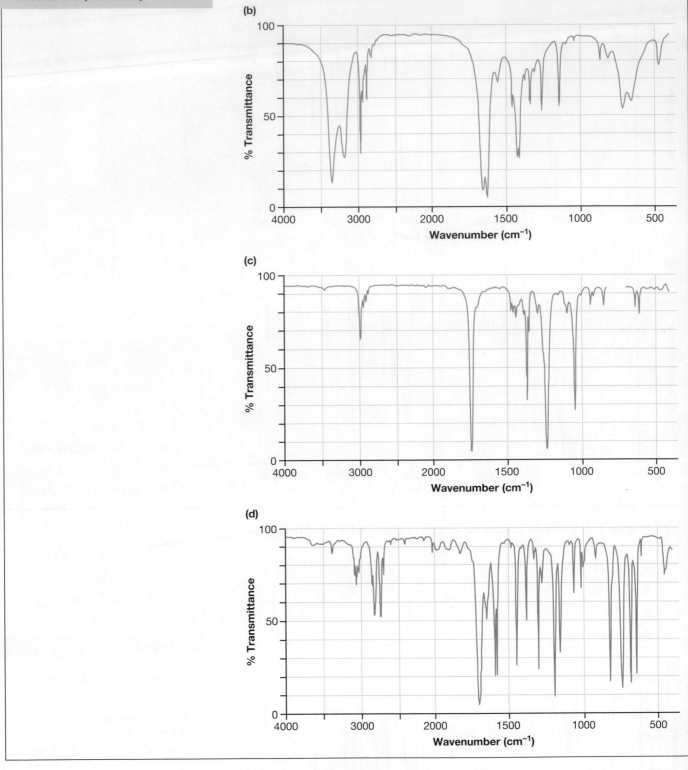

NMR Spectroscopy

CAN YOU PICTURE A BELL FROM ITS RING?

The ringing of bells produces characteristic sounds that can be related to their shapes and sizes. We can "listen" to information about shapes and sizes of molecules through NMR spectroscopy.

Top: Alexander Blinov/Alamy Stock Photo. Left: Stefan Roller/ Shutterstock. Right: Daan Kloeg/Shutterstock.

Nuclear magnetic resonance (NMR) spectroscopy is one of the most important diagnostic tools available to an organic chemist. It provides a spectrum of peaks containing information on the number of protons (^{1}H) and carbons (^{13}C), as well as their proximity to electronegative atoms. The NMR spectrum also indicates how structural features are connected, even at locations distant from the functional groups. The infrared (IR) spectrum is a very sensitive indicator of the types of functional groups, but provides less information about connectivity. As a result, NMR and IR work synergistically to determine the structure of an organic compound. In many cases, NMR can be used to determine a complete chemical structure in a very short period of time. NMR can also define product ratios, measure purity, and identify impurities.

7.1 ^{1}H NMR SPECTROSCOPY

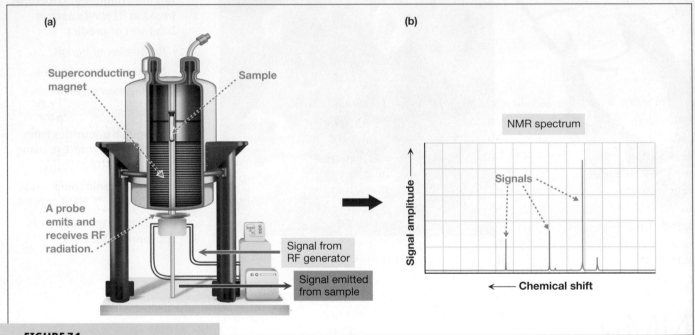

FIGURE 7.1

(a) A typical high-field nuclear magnetic resonance (NMR) spectrometer and (b) the output NMR spectrum.

Karty, J. *Organic Chemistry: Principles and Mechanisms*, 3rd ed.; W. W. Norton: New York, 2022; p 820.

magnetic dipole transitions >>
A change in the spin state of a magnetically active nucleus in response to absorption or emission of radiofrequency energy.

7.1A Theory of ^{1}H NMR Spectroscopy

Radiofrequency radiation is the part of the electromagnetic spectrum used for NMR spectroscopy (Figure 6.1). When a sample is placed into the magnetic field of a large superconducting magnet (**Figure 7.1**), and radiofrequency energy is used to monitor the **magnetic dipole transitions** of the nuclei, a spectrum is obtained that can be used to determine the structures of the compounds in the sample. Many labs use smaller, inexpensive, low-field instruments that operate with permanent magnets; although these furnish spectra with lower resolution, they are suitable for basic applications.

The spinning charge of a **spin-active nucleus** creates a small magnetic dipole, called μ (**Figure 7.2**). When the nucleus is under the influence of a much larger external magnetic field H_0, then μ has a tendency to align with H_0 in one of two **spin states**. The magnetic dipole μ of the nuclear spin will be either aligned in the same orientation as H_0 (the lower energy α state) or aligned in the opposite orientation (the higher β energy state).

The energy difference ΔE between these α and β states corresponds to the amount of energy imparted by absorbing a photon in the radiofrequency range of the spectrum. When the frequency of irradiation is matched, or in *resonance* with ΔE, it can be absorbed, and this absorption causes an excitation from the α state to the higher energy β state (**Figure 7.3**). The spin is said to "flip" from α to β. Over a short period of time (usually within a couple of seconds), the excited state β undergoes relaxation, back to the lower energy α state, and energy is emitted.

<< **spin-active (or magnetically active) nuclei**
Nuclei that exhibit nuclear magnetic resonance, and can give signals in NMR spectroscopy. Generally this refers to nuclei with a nonzero nuclear spin quantum number (I). Nuclei of ^1H and ^{13}C have $I = 1/2$, are magnetically active, and give NMR signals, while the ^{12}C nucleus has $I = 0$ and is not spin-active.

<< **spin states**
Different spin energy levels in which the magnetic dipole of a nucleus is aligned with or against an external magnetic field.

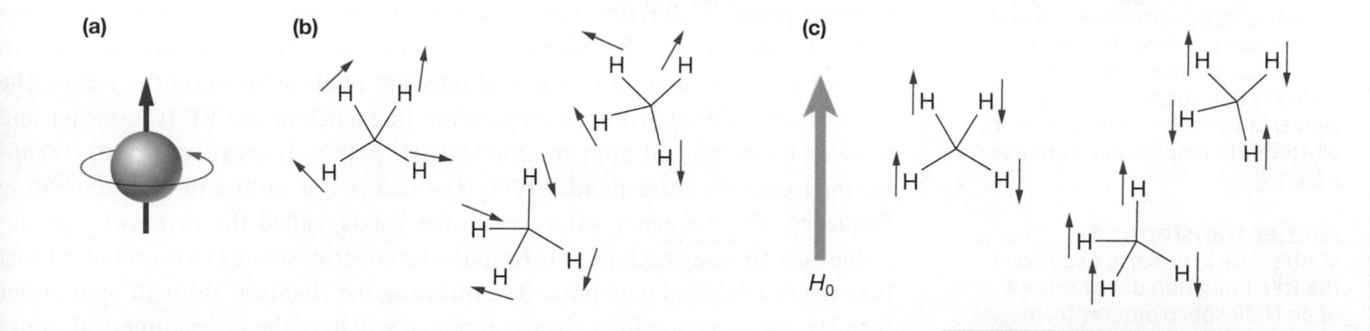

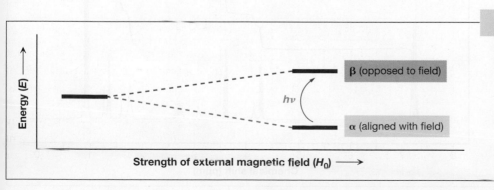

FIGURE 7.2

(a) A spinning nucleus creates a magnetic dipole μ, indicated by the solid arrow. (b) A sample of methane (CH_4) in the absence of any external magnetic field; the magnetic dipole of each hydrogen nucleus is randomly oriented. (c) A sample of methane in a strong external magnetic field (H_0); the magnetic dipoles of each hydrogen are aligned with (α) or opposing (β) the external field, with a slight excess in the lower energy α orientation.

The sample contains a population of nuclei distributed between the α and β spin states, with the lower energy state slightly in excess (**Figure 7.4**). Absorption of radiofrequency (RF) energy causes an increase in the population of the β state. The net magnetization associated with the population of dipoles changes as a result. As the population relaxes back to its initial distribution, and the net magnetization returns to its prior equilibrium condition, small fluctuations in the magnetic field in the sample create a current in the radiofrequency receiver, generating the NMR signal. The signal strength is low, so the NMR spectrum is obtained using a series of RF pulses, with a short delay after each pulse. With numerous repetitions, the signals become magnified and random noise is eliminated, improving the quality of the spectrum.

Nuclear magnetic resonance is extremely sensitive to small differences in the environment around the nucleus, including other nearby nuclei and their associated electrons. When several nuclei are present in different parts of a molecule, their emissions occur at slightly different frequencies. The emitted signals are all

FIGURE 7.3

The difference in energy ($\Delta E = h\nu$) between α and β states is proportional to the applied external magnetic field (H_0). Absorption of a radiofrequency photon of energy $h\nu$ causes the magnetic dipole μ to transition from the α energy level to the β energy level.

Energy (E)

$h\nu$

β (opposed to field)

α (aligned with field)

Strength of external magnetic field (H_0) →

FIGURE 7.4

Changes in population of nuclear spins in α and β states upon irradiation. The frequency ν is in the radiofrequency (RF) region of the electromagnetic spectrum, associated with a change of the net magnetization in the sample. The relaxation back to the original population distribution results in emission of energy.

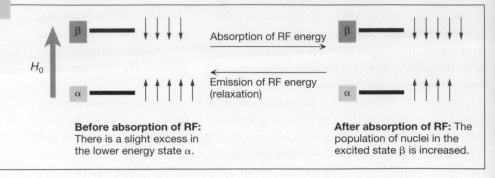

Before absorption of RF: There is a slight excess in the lower energy state α.

After absorption of RF: The population of nuclei in the excited state β is increased.

free induction decay (FID) >>
Raw form of radiofrequency emission signals from nuclei relaxing from a higher energy spin state to a lower energy spin state, plotted with respect to time. This output from a Fourier transform NMR instrument is mathematically converted into the NMR spectrum, plotted with respect to frequency (Hz).

Fourier transform >>
Mathematical process to convert the free induction decay output of an NMR spectrometer from the time domain to the frequency domain, resulting in the NMR spectrum.

contained together in a data set called a **free induction decay (FID)**, and to make use of this information, it is processed by a computer using **Fourier transform** operations to translate the emissions into a spectrum (**Figure 7.5**). The spectrum contains peaks in different positions that correspond to different nuclei in the molecular structure of the sample, and is plotted with frequency on the x axis, in units of parts per million (ppm), and intensity of absorbance on the y axis. The units of ppm are relative to the operating frequency of the RF transmitter and receiver; for example, 1 ppm at a 300 MHz (1 MHz = 1 megahertz = 10^6 Hz) operating frequency corresponds to 300 Hz—that is, one millionth of the operating frequency. The frequency values along the x axis, called the *chemical shift*, describe how far away each peak is from the tetramethylsilane (TMS) standard that is assigned a value of 0.00 ppm. By expressing the chemical shifts in ppm rather than Hz, the spectrum for a given compound will have the same numerical values on the x axis regardless of the operating frequency of the instrument. Each peak offers information about the molecular structure surrounding each of the nuclei, as judged by the chemical shift and other information contained in the peaks.

Nuclei with nonzero nuclear spin can be examined by NMR. When there is an even number of protons and also an even number of neutrons in the nucleus, the nuclear spin $I = 0$. These nuclei give no NMR signal; ^{12}C (6 protons, 6 neutrons) and ^{16}O (8 protons, 8 neutrons) are common examples. ^1H (1 proton, 0 neutrons) and ^{13}C (6 protons, 7 neutrons) are the most common nuclei that organic chemists subject

FIGURE 7.5

Example of a ^1H NMR spectrum from a research-grade high-field instrument with a 300 MHz operating frequency. Each vertical rise of the blue line measures the area, or integration, of the corresponding peak.

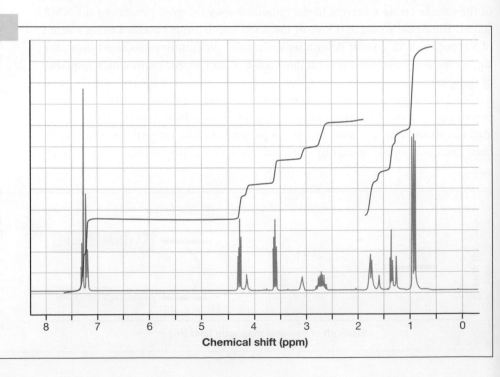

Chemical shift (ppm)

to NMR spectroscopy. The combination of protons and neutrons is an odd number for these nuclei, and each has $I = 1/2$. These nuclei are present throughout organic structures, not just in the functional groups. For that reason, NMR makes it possible to glean detailed information about portions of molecular structures independent of functional groups—information that is usually unavailable from UV–vis or IR spectroscopies.

Sample Problem

Two peaks appear 0.20 ppm apart in a spectrum obtained with an instrument operating at a frequency of 400 MHz. How far apart are they in Hz?

Solution

At an operating frequency of 400 MHz, one millionth of that, or one ppm, is 400 Hz. Use this relationship of 400 Hz per ppm to convert ppm to Hz:

$$\Delta\delta\,(\text{Hz}) = 0.20\ \text{ppm} \times \frac{400\ \text{Hz}}{1.00\ \text{ppm}}$$
$$= 80\ \text{Hz}$$

7.1B Interpreting Data in the NMR Spectrum

There are four main types of information available from a ^{1}H NMR spectrum. Each of these correlates to specific structural details, as follows:

1. Number of signals: The number of distinct signals corresponds to the number of different types of H, which is related to molecular symmetry.

2. Chemical shift: The frequencies of peaks (i.e., their location on the x axis of the spectrum) indicate electronic effects of nearby atoms and functional groups.

3. Integration: The area under each peak indicates how many H's give rise to that peak.

4. Multiplicity: A peak may have finer details because the signal is split into more than one part; the splitting indicates how many H's are present on neighboring atoms and may also reveal information about how they are spatially arranged relative to each other.

In ^{13}C NMR spectroscopy, the number of ^{13}C signals and their chemical shifts are the key elements to evaluate. The spectrum is acquired and processed in such a way that integration and multiplicity are not generally useful. We return to ^{13}C NMR in more detail near the end of this chapter.

In the next four sections, we explain in detail how to find and use all four types of information from ^{1}H NMR spectra. For a careful analysis of the NMR spectrum, especially when the identity of the compound is unknown, all four can be evaluated. However, not all of these types of information may be necessary to solve every problem. If two possible structures have a different predicted number of signals, for example, then one of them may be ruled out simply by counting the signals in the observed spectrum.

7.1C Number of Signals: Equivalent and Nonequivalent Hydrogens

The number of signals, or peaks, in a spectrum corresponds to the number of nonequivalent hydrogens. If hydrogens are nonequivalent, they will have different chemical and magnetic environments that depend on nearby structural features.

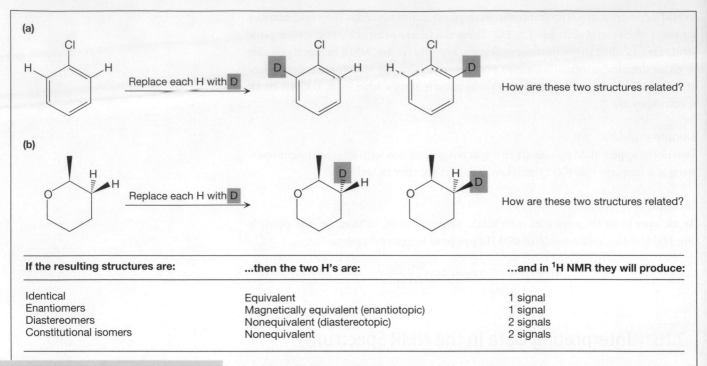

(a)

Replace each H with D

How are these two structures related?

(b)

Replace each H with D

How are these two structures related?

If the resulting structures are:	...then the two H's are:	...and in ¹H NMR they will produce:
Identical	Equivalent	1 signal
Enantiomers	Magnetically equivalent (enantiotopic)	1 signal
Diastereomers	Nonequivalent (diastereotopic)	2 signals
Constitutional isomers	Nonequivalent	2 signals

FIGURE 7.6

Depiction of how to determine whether two hydrogens will appear in the same signal or in different signals. Sequentially replace each by another atom, then determine the relationship between the resulting structures. *Solutions:* (a) Identical, 1 signal. (b) Nonequivalent (diastereotopic), 2 signals.

This means they will have different resonance energies, and this will give a peak in the spectrum at different frequencies. Equivalent hydrogens, on the other hand, all appear in the same signal. This arises when there is symmetry in a structure, such as an internal plane of symmetry that makes the left and right sides of a structure identical, or a symmetry axis such as the rotation about a C—C bond that makes all the H's of a methyl group equivalent.

Sometimes a symmetry plane or symmetry axis may not be obvious, so there is a simple test that can help determine if any two hydrogens in a structure are equivalent or nonequivalent (**Figure 7.6**). First, draw two copies of the structure, including the two hydrogens you want to test. On one structure, replace one of the two hydrogens with another atom (such as D, deuterium), and on the second structure make the same replacement of the other hydrogen. Now, determine how these two structures are related. Are they identical or different? Are they constitutional isomers or stereoisomers? If they are stereoisomers, are they enantiomers or diastereomers?

If structures A and B are

- *identical*, then the original two hydrogens before replacement will be equivalent, and these H's will appear in the same signal in the NMR spectrum.
- *constitutional isomers*, then the original two hydrogens before replacement will be nonequivalent, and these H's will have different peaks in the NMR.
- *enantiomers*, then the original two hydrogens before replacement will be *enantiotopic*, and these H's will be equivalent in the NMR.
- *diastereomers*, then the original two hydrogens before replacement will be *diastereotopic*, and these H's will be nonequivalent in the NMR.

To remember these last two, keep in mind that enantiomers have the same physical and spectroscopic properties (except optical rotation), but diastereomers have different properties—including nuclear resonance energy and other properties related to spectroscopy.

Some examples of structures that offer one or more of these scenarios are presented in Figure 7.6. Predict how many signals should appear in each ¹H NMR spectrum.

Certain types of hydrogens that are involved in hydrogen bonding, specifically O—H, N—H, and S—H, may not be detected reliably in all conditions. Hydrogen bonding interactions between molecules containing these bonds can lead to exchange of the hydrogens between molecules in a process known as **chemical exchange of hydrogen**. When the exchange is fast, the H's will be moving around among many different magnetic environments, and the signal in an NMR spectrum will be an average of all those different environments. This can lead to a signal that is broad, and may appear at chemical shifts that vary with concentration and other sample conditions. With carboxylic acids, the hydrogen of the CO_2H group can be so broad that you might not see it clearly as a peak, though it may be obvious in the integration.

Chemical exchange can help identify which peaks in a spectrum belong to exchangeable hydrogens—that is, those that are attached to electronegative atoms. **Deuterated NMR solvents** that are hydrogen-bond donors, such as CD_3OD or D_2O, can exchange their deuterium atoms for the exchangeable hydrogen atoms in the sample. The ^1H NMR spectra acquired in these solvents, therefore, will *not* show any signals for H's that are in O—H, N—H, or S—H bonds. To facilitate comparison with a reference spectrum acquired in $CDCl_3$, a few drops of D_2O can be shaken with the $CDCl_3$ solution in the NMR tube, and a new spectrum acquired. Comparing the spectra acquired before and after deuterium exchange makes it possible to assign the peaks that disappeared to the exchangeable hydrogens in the structure.

<< **chemical exchange of hydrogen**
Rapid proton transfer reactions between molecules containing hydrogens attached to electronegative atoms. This is commonly observed with hydrogen bond donor groups such as OH, NH, and SH.

<< **deuterated NMR solvent**
A solvent that has had its ^1H atoms replaced by ^2H, or deuterium (D) atoms that do not absorb radio-frequency energy in the same frequency range as ^1H. The solvent properties remain the same, but the deuterated solvent does not appear in a ^1H NMR spectrum. Trace amounts of the ^1H solvent are generally still present, resulting in a very small solvent peak; this can be used for calibration of the spectrum.

7.1D Chemical Shifts

The frequencies of the peaks (in units of ppm) in the NMR spectrum can be correlated to structural features, and this allows us to distinguish between hydrogen nuclei that are in different environments. There is a significant difference in the frequency of the peaks associated with the methyl groups of dimethyl ether (CH_3OCH_3, 3.2 ppm), and dimethyl sulfide (CH_3SCH_3, 2.1 ppm). Once we know the frequency of absorption for these two compounds, we can use the NMR spectrum of a laboratory sample to identify whether one or both of these compounds may be present. Frequencies of ^1H NMR peaks are known for millions of organic compounds, and when a new compound is made, the frequencies are measured and recorded in journals and databases so that other chemists can use the data for comparison.

Rarely does a spectrum have only one peak at one frequency, however. A more complicated molecule—one with lots of different types of hydrogens—affords peaks at a variety of frequencies. As first mentioned when we introduced the theory of ^1H NMR spectroscopy, NMR data are generally presented in a plot where the y axis is intensity of absorbance and the x axis is chemical shift (δ), measured in units of ppm (parts per million). ΔE (or v) increases from right to left. These chemical shifts vary slightly depending on the environment around the hydrogens (i.e., what functional groups are nearby). Certain chemical shift ranges in the spectrum correlate very well to specific types of functional group environments (**Figure 7.7**).

The electronegativity of neighboring atoms dramatically affects the chemical shifts. The electrons around the nucleus cause magnetic **shielding**, which counteracts the external field H_0. Nearby electronegative atoms have an electron-withdrawing effect that lessens the shielding; this effect is called **deshielding**. Shielded hydrogens appear to the right, or *upfield*, while deshielded hydrogens appear at the left of the spectrum, or *downfield*. Figure 7.7 shows that a hydrogen in the absence of any functional group, electronegative atom Z, or π bond appears at about 1 ppm. One electronegative atom, such as a halogen or oxygen, shifts the resonance frequency downfield by about 2 ppm. These electronegative Z groups have an additive effect on the deshielding. For example, the hydrogens of CH_2Cl_2 give a peak at 5.28 ppm, and the hydrogen of $CHCl_3$ appears at 7.26 ppm. The chlorine atoms are diminishing the electron density around

<< **shielding/deshielding**
The effect on a nucleus from nearby electron density that changes the frequency of radio-frequency absorption. A nearby electronegative atom reduces electron density around a nucleus, resulting in deshielding.

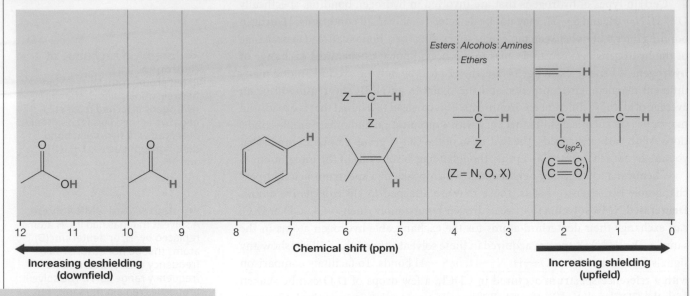

FIGURE 7.7

How chemical shift correlates with different functional group environments.

Chemical shift (ppm)

Increasing deshielding (downfield)

Increasing shielding (upfield)

Esters Alcohols Amines
Ethers

(Z = N, O, X)

$C_{(sp^2)}$
$\begin{pmatrix} C=C, \\ C=O \end{pmatrix}$

the H's, so there is less of a shielding effect. Deshielding increases the energy difference between the α and β spin states (**Figure 7.8**), causing the emission to be of higher energy and farther downfield.

It is not only electronegativity that can cause deshielding. The benzene ring causes hydrogens to appear downfield, even though it contains no electronegative atoms. This is due to a phenomenon called **magnetic anisotropy**, which is caused by circulating electrons in π systems (**Figure 7.9**). The circulating π electrons create a current, inducing another magnetic field that reinforces the applied magnetic field, H_0, in the location of the attached hydrogens. Due to anisotropy, protons near double bonds or aromatics are shifted downfield (to the left in the spectrum) in comparison with an alkane H (**Figure 7.10**). Protons directly attached to aromatics or double bonds are shifted the most. The downfield shift of terminal alkyne H is less dramatic, although triple bonds do exhibit anisotropy.

(a) Shielded   **(b) Deshielded**

β β

ΔE ΔE

α α

FIGURE 7.8

Effects on the relative energy levels of the α and β states for a hydrogen nucleus that is either (a) shielded or (b) deshielded by an electronegative atom Z.

magnetic anisotropy »
An induced magnetic field that can be additive to or subtractive from the applied field experienced by a nucleus, depending on where the nucleus is located relative to the source of the induced field. This is commonly observed with ^1H nuclei located near aromatic rings and other π systems.

7.1E   Integration

A tool from calculus, the integral, gives us another piece of information about the peaks in our NMR spectrum. Integration is a measurement of the area under a curve, and this can be used to measure the area under a peak in an NMR spectrum. In NMR spectroscopy, the integral of a peak is proportional to the number of hydrogens giving rise to that peak. The units of integration don't matter, because they are relative values used to compare two peaks. As long as the units are the same for both peaks, the ratio of the integrals for each peak will be useful. So, whether the integral ratio is 120:40 or 7.5:2.5, the ratio is the same (3:1), and it corresponds to the ratio of the number of hydrogens giving rise to the two peaks. Depending on how the operator formats the printed spectrum, the integral measurements may be illustrated graphically, with a line inscribed over the spectrum (the blue line in **Figure 7.11**) or the numerical values may be printed on the spectrum.

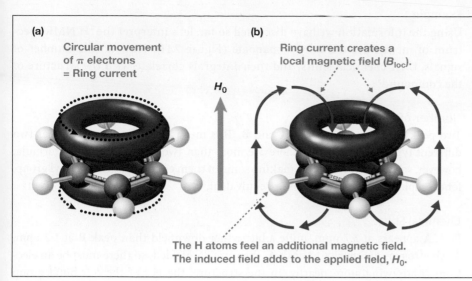

(a)

Circular movement
of π electrons
= Ring current

H_0

(b)

Ring current creates a
local magnetic field (B_{loc}).

The H atoms feel an additional magnetic field.
The induced field adds to the applied field, H_0.

FIGURE 7.9

Anisotropy effects upon chemical shift of hydrogens located near the π system of benzene.

Karty, J. *Organic Chemistry: Principles and Mechanisms*, 3rd ed.; W. W. Norton: New York, 2022; p 833.

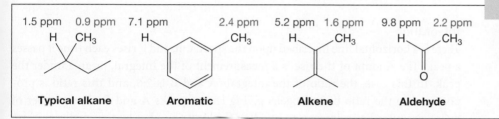

1.5 ppm   0.9 ppm   7.1 ppm   2.4 ppm   5.2 ppm   1.6 ppm   9.8 ppm   2.2 ppm

Typical alkane   **Aromatic**   **Alkene**   **Aldehyde**

FIGURE 7.10

Typical chemical shifts of H and CH_3 attached to the sp^2-hybridized carbons of an aromatic ring, alkene, and aldehyde.

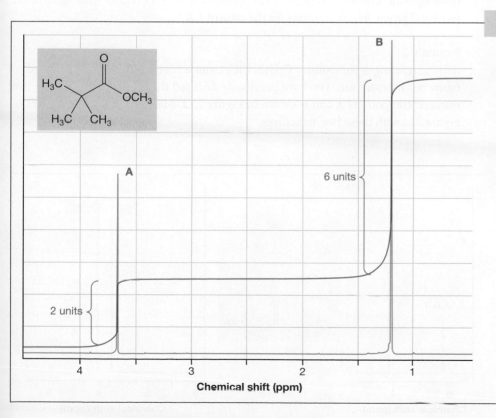

FIGURE 7.11

^{1}H NMR spectrum of methyl 2,2-dimethylpropanoate (300 MHz, CDCl$_3$). The integration of the spectrum is shown with the blue line. Gridlines are added to the spectrum; in this example, each horizontal gridline is one unit to measure the rise in the integral line. Any unit (mm, cm, etc.) may be used, as long as it is consistent for all peaks within the spectrum.

© Sigma-Aldrich Co. LLC. Reproduced with permission from Merck KGaA, Darmstadt, Germany and/or its affiliates.

Example 1

Using the information we have discussed so far, let's interpret the ^{1}H NMR spectrum of methyl 2,2-dimethylpropanoate (Figure 7.11). How do the number of signals, their chemical shifts, and their integrals correlate with the structure of the compound?

Number of Signals

Two peaks are present, labeled **A** and **B**. This means that the compound has two different types of hydrogens. There are more than two hydrogens in the formula, however, so symmetry must be making some of them equivalent. In a methyl group, for example, all three hydrogens are equivalent.

Chemical Shift

Peak **A** appears at 3.7 ppm, quite a bit farther downfield than peak **B** at 1.2 ppm. The hydrogens giving rise to peak **A**, then, are deshielded, so there must be an electron-withdrawing atom nearby. In the structure, the H's of the *tert*-butyl group are four bonds away from the nearest electronegative atom (either oxygen atom), whereas the H's of the methoxy group are only two bonds away, and the methoxy group would therefore experience deshielding.

Integration

There is a horizontal line inscribed upon the spectrum, and it rises each time it passes a peak. The amount of the rise is a measurement of the integral, or area under the peak. In this case, the ratio of the integrals **A** and **B** is 2:6, and this ratio is proportional to the ratio of hydrogens giving rise to peaks **A** and **B**. The number of hydrogens must be an integer, so dividing the larger number by the smaller one, we get a ratio of 1:3. According to the structure (Figure 7.11), there are 12 hydrogens—three equivalent hydrogens in the OCH_3 group and nine equivalent hydrogens in the *tert*-butyl group, which accounts for the ratio of 1:3.

Example 2

After extracting a compound, C_3H_4BrClO, from a marine organism, you need to figure out its structure. You have previously deduced that you have one of the two isomeric compounds **A** and **B** shown in **Figure 7.12**. Match the ^{1}H NMR spectra in Figure 7.12 with these two structures.

FIGURE 7.12

Two predicted ^{1}H NMR spectra for two isomers of C_3H_4BrClO. The integral ratios have been calculated, and are shown at the top of each peak.

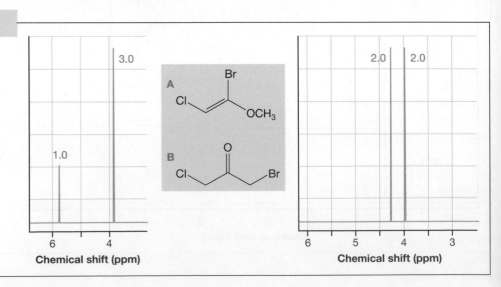

First narrow down the possible structures using the molecular formula along with functional group data from infrared or NMR (for this problem, the structures are already provided in Figure 7.12). Then, for each structural hypothesis, predict the number of signals and their relative integrals. Finally, compare these predictions with the observed spectrum, ruling out any structures that are inconsistent with observations.

Each of the structures has two different types of hydrogen, so each would be expected to have two signals in the NMR spectrum. Each of the spectra in Figure 7.12 has two signals, so we cannot differentiate them by number of signals. Structure **A** has an alkene, and a hydrogen attached to an alkene generally appears at 4.5–6.5 ppm (see Figure 7.7). Only the spectrum on the left has a peak that matches this expected chemical shift, so we might conclude that that spectrum corresponds to structure **A**. Can we strengthen this conclusion by analyzing the integration? Structure **A** would be expected to have two peaks, one for a single alkene hydrogen and the other for a methyl group (3 hydrogens); the two peaks would have an integral ratio of 1:3. This also matches the spectrum on the left, and strengthens our conclusion because two independent pieces of evidence both lead to the same conclusion.

The spectrum on the right in Figure 7.12 has two peaks with an integral ratio of 2:2, the ratio we would expect for structure **B**. How do we know which hydrogens belong to which signal? Recall that nearby electronegative atoms deshield the nearby H's, shifting their signals to the left (downfield). Because Cl is more electronegative than Br, the hydrogens nearer the Cl should be more deshielded, and farther downfield.

7.1F | Multiplicity (Signal Splitting)

In the NMR spectra we have seen so far, there has been only one line, or one maximum on the y axis, within each signal. Each of these peaks, called a *singlet*, is what we observe for hydrogens that are several bonds away from any other nonequivalent hydrogens. Commonly, there are nonequivalent hydrogens located in a **vicinal** (3 bonds away) or **geminal** (2 bonds away) relationship, and in these cases there will be **signal splitting**. Instead of a singlet, the peak will appear to have more than one maximum within it—it has been split into a *multiplet*. The number of maxima within a multiplet is called the **multiplicity**, and this can be correlated with the number of nonequivalent H's that are nearby. Multiplicity is therefore a powerful tool to unravel how the atoms are connected within the structure.

Signal splitting occurs when nonequivalent hydrogens are within two or three bonds of each other because their spins are coupled. They are close enough that the magnetic dipole of one H can add to or subtract from the overall magnetic field of the other. If the hydrogens are four bonds apart or more, then their magnetic dipoles have very little impact on each other, and usually no splitting is observed. The strong external field H_0 is accompanied by the small magnetic dipoles associated with nearby hydrogen nuclei. If a nearby dipole is aligned with H_0, it adds to H_0, whereas if it is aligned opposite H_0, it subtracts from H_0. This, in turn, affects the frequency of the absorbance. Thus, one nearby hydrogen can split a signal into two signals, because its dipole has two possible alignments (i.e., with or against H_0).

The presence or absence of signal splitting is associated with particular structural features (**Figure 7.13**). Most commonly, signal splitting is observed with vicinal hydrogens (neighboring hydrogens, on two carbons that are attached to each other) and in order to observe it, the vicinal hydrogens must be nonequivalent. Normally geminal hydrogens (hydrogens attached to the same carbon) are equivalent, though

<< **vicinal**
The relationship between atoms that are separated by three bonds.

<< **geminal**
The relationship between atoms that are separated by two bonds.

<< **signal splitting**
Creation of two or more lines from one NMR signal caused by induced magnetic fields from nearby nuclear spins (usually ^1H), which can be additive to or subtractive from the applied field, depending on whether the nearby nuclear spin is aligned with or against the applied field. The effect is most common for ^1H nuclei that are 2 or 3 bonds apart, but longer-range splitting is sometimes observed.

<< **multiplicity**
The number of lines within a signal when signal splitting occurs due to neighboring nuclei, usually ^1H.

FIGURE 7.13

Nearby hydrogens may or may not give rise to signal splitting.

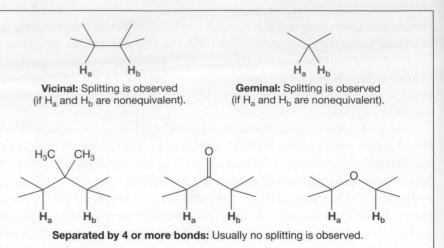

Vicinal: Splitting is observed (if H$_a$ and H$_b$ are nonequivalent).

Geminal: Splitting is observed (if H$_a$ and H$_b$ are nonequivalent).

Separated by 4 or more bonds: Usually no splitting is observed.

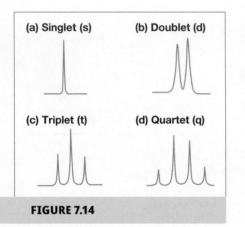

(a) Singlet (s) **(b) Doublet (d)**

(c) Triplet (t) **(d) Quartet (q)**

FIGURE 7.14

they can show splitting if they are nonequivalent. There is no splitting of signals in Figure 7.12, where the geminal H's in structure **B** are equivalent. Geminal splitting can sometimes be seen in cases where cis–trans isomerism is possible, such as a CH$_2$ of a substituted cyclohexane. One H is cis to a substituent, and the other is trans, so they are diastereotopic, and nonequivalent.

Coupling and signal splitting can be described by two simple rules. First, to observe coupling, the vicinal hydrogens must be in different environments (nonequivalent). Second, the splitting occurs according to the $N + 1$ rule. When there are N equivalent hydrogens nearby (vicinal, usually), they will split the signal into $N + 1$ peaks:

- No vicinal hydrogens: Signal appears as a single peak, called a *singlet* (**Figure 7.14a**; $N = 0$; $N + 1 = 1$).
- One vicinal hydrogen: Signal appears as two peaks, called a *doublet* (**Figure 7.14b**; $N = 1$; $N + 1 = 2$).
- Two vicinal hydrogens: Signal appears as three peaks, called a *triplet* (**Figure 7.14c**; $N = 2$; $N + 1 = 3$).

The five spectra in **Figure 7.15** illustrate these features. In spectrum (a), the two vicinal hydrogens are equivalent, so no coupling is observed. In spectrum (b), the two vicinal hydrogens are nonequivalent because of the different environments around Cl and Br. This gives rise to two different signals, and each of the two hydrogens splits the other into a doublet.

FIGURE 7.15

Examples of vicinal coupling that leads to signal splitting.

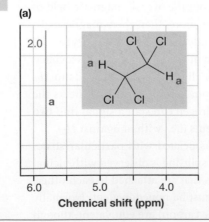

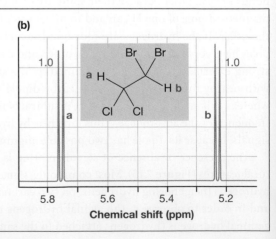

(continued)

FIGURE 7.15 *(continued)*

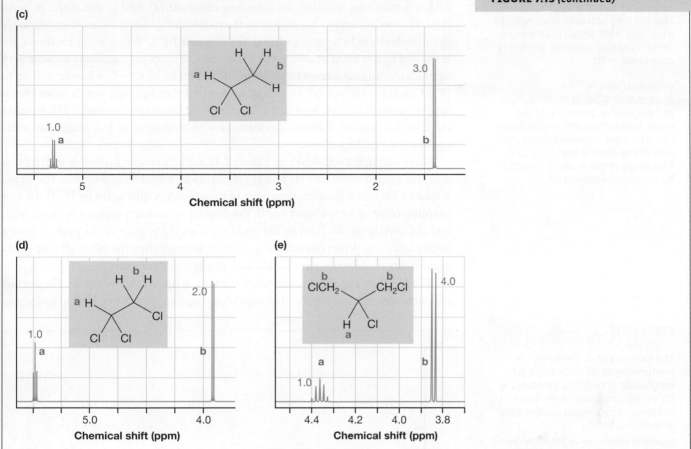

In spectrum (c), H_a is split into a *quartet* (**Figure 7.14d**), because there are three vicinal hydrogens in the neighboring methyl group ($N = 3$, $N + 1 = 4$); its signal has four peaks within it, and they are of unequal intensity. Integrating the area under each part of the peak reveals that the ratio of the four peaks within the quartet is 1:3:3:1, which is typical for a quartet. Similarly, the triplet in spectrum (d) consists of three peaks of unequal intensity in a ratio of 1:2:1. This is typical of a triplet. In both spectra (c) and (d), H_a splits the signal for the H_b hydrogens into a doublet.

In spectrum (e), there are two sets of hydrogens labeled H_b on different carbons, but because of the symmetry in the structure, all four of them are equivalent. Thus, H_a is split into a signal of multiplicity 5 ($N = 4$, $N + 1 = 5$), or a quintet. The signal for H_b is split only by one neighbor (H_a) and is observed as a doublet. What will be the area ratios of these peaks within the quintet? You can predict the ratios by using Pascal's triangle (**Figure 7.16**). Simply put, Pascal's triangle gives you the area ratios for all multiplets, from singlets to doublets to triplets to quartets to quintets to sextets to septets and so on. In the chemical literature, once you get beyond quartets, the higher multiplets are often just referred to as multiplets.

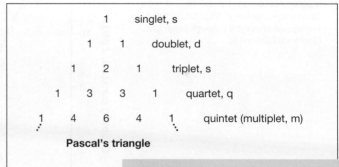

Pascal's triangle

FIGURE 7.16

Pascal's triangle predicts the relative peak intensity in multiplets.

COUPLING TO MORE THAN ONE TYPE OF HYDROGEN

If two different types of vicinal hydrogens split a signal, then more complicated splitting patterns can result, because the magnitude of the splitting is not always equal. The spacing between the peaks in multiplets is not necessarily the same from one molecule to the next, or in different parts of a molecule. This spacing within a multiplet is called the **coupling constant** (J), and is measured in hertz (Hz). The variations in J are related to the molecular geometry—most importantly the **dihedral angle** between vicinal H's (**Figure 7.17**). When bond rotations are restricted by a π bond or a ring, the differences in coupling constants become very significant, ranging from 0 to 10 Hz for those linked by C—C σ bonds, and up to 18 Hz for trans vicinal H's on an alkene. Freely rotating single bonds in saturated acyclic molecules often lead to vicinal coupling constants around 7 Hz, because the contributions of different conformers with various dihedral angles are averaged by the rapid rotations.

Some examples are shown in **Figure 7.18** for a hydrogen H_a that is split by two different vicinal neighbors H_b and H_c, resulting in a *doublet of doublets*. The signal is split by H_b into a doublet, then each of those lines is split again by H_c. If the two coupling constants are almost equal, the doublet of doublets appears to be a triplet, and the overlap of two lines in the middle causes the middle of the peak to have a higher intensity. When one coupling constant is larger than the other, all four peaks of the doublet of doublets are visible with nearly equal intensity.

The $N + 1$ rule applies to cases when there are N *equivalent* hydrogens. The doublet of doublet examples in Figure 7.18 occur when there are two *nonequivalent* hydrogens

FIGURE 7.17

The Karplus curve, depicting the relationship of dihedral angle ϕ to magnitude of coupling constant J for vicinal couplings (three-bond coupling to hydrogens attached on adjacent carbons).

Used with permission of The American Institute of Physics. Karplus, M. Contact Electron-Spin Coupling of Nuclear Magnetic Moments. *J. Chem. Phys.* © **1958**, *30* (1). https://doi.org/10.1063/1.1729860. Permission conveyed through Copyright Clearance Center, Inc.

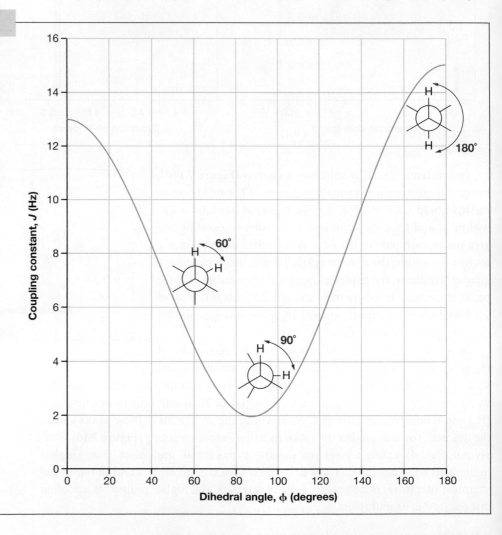

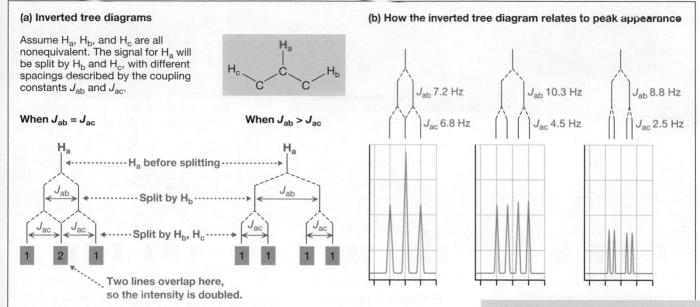

(a) Inverted tree diagrams

Assume H_a, H_b, and H_c are all nonequivalent. The signal for H_a will be split by H_b and H_c, with different spacings described by the coupling constants J_{ab} and J_{ac}.

When $J_{ab} = J_{ac}$

H_a

H_a before splitting

J_{ab} — Split by H_b

J_{ac} J_{ac} — Split by H_b, H_c

1   2   1

Two lines overlap here, so the intensity is doubled.

When $J_{ab} > J_{ac}$

H_a

J_{ab}

J_{ac} J_{ac}

1   1   1   1

(b) How the inverted tree diagram relates to peak appearance

J_{ab} 7.2 Hz
J_{ac} 6.8 Hz

J_{ab} 10.3 Hz
J_{ac} 4.5 Hz

J_{ab} 8.8 Hz
J_{ac} 2.5 Hz

FIGURE 7.18

Coupling of nonequivalent hydrogens H_b and H_c to H_a, when the coupling constants J are different, leads to a *doublet of doublets*.
(a) An inverted tree diagram helps to describe how the signal of H_a is split, first by H_b, then by H_c.
(b) The appearance of the doublet of doublets varies depending on the relative values of the coupling constants.

nmrdb.org: Tools for NMR Spectroscopists [Online]. https://www.nmrdb.org/ (accessed April 2022). CC-BY-4.0 https://creativecommons.org/licenses /by/4.0/

splitting the signal. In this latter case, instead of the $N + 1$ rule, there is an expanded form of the $N + 1$ rule that treats the two different hydrogens separately:

$$\text{Multiplicity} = (N + 1)(M + 1)$$

The effect of one of the hydrogens N is described by the $N + 1$ rule, and we introduce a label M for the different set of hydrogens and handle its effect similarly; then we multiply the two effects together. With two different hydrogens splitting the signal, both N and M are 1, and we have

$$\text{Multiplicity} = (1 + 1)(1 + 1) = 4$$

The signal in such a case will have four separate lines, and we can see this in the doublet of doublet cases (Figure 7.18). Further examples of multiplicity that give rise to doublets of triplets or even more complicated patterns are shown in **Figure 7.19**. You'll rarely encounter these more complex situations in introductory organic laboratory courses, but they are commonly seen in more advanced courses and in research labs.

COUPLING IN ALKENES

For alkenes, there is no rotation of the π bond under typical conditions, and this means that the dihedral angles for vicinal H's attached to the planar alkene are fixed, varying only slightly from the values of either 0° or 180°. The cis H's have a dihedral angle of 0° and coupling constants around 8–10 Hz, whereas trans H's have a dihedral angle of 180° and coupling constants around 15–17 Hz. As a result, it is possible to assign cis or trans configuration using the coupling constants, especially if the spectrum of both isomers is available for comparison. Also, geminal coupling may be observed in alkenes when there are two nonequivalent H's attached at one end of an alkene. Alkene geminal coupling is typically quite small, 0–2 Hz, so it may not clearly show up in the spectrum without an expansion plot of the peak in question.

For *n*-butyl vinyl ether (**Figure 7.20**), the alkene region of the spectrum shows coupling constants of 14.3 Hz, 6.8 Hz, and 1.8 Hz. The peaks at 4.2 and 3.9 ppm, assigned to H_b and H_c, respectively, are not simple doublets, as would be expected for vicinal coupling; instead, they show a second smaller coupling constant that is attributed to geminal coupling.

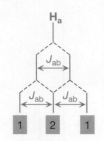

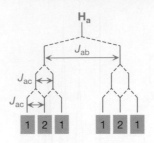

Splitting of H_a, by two **nonequivalent** sets of H's ($N = 1$ and $M = 2$). Coupling constants (J_{ab} and J_{ac}) are different:

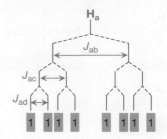

Splitting of H_a, by three **nonequivalent** sets of H's ($N = 1$, $M = 1$, $P = 1$). Coupling constants (J_{ab}, J_{ac}, and J_{ad}) are different:

(a)

H_a

Multiplicity = $(N + 1) = 3$

Peak splitting pattern:
triplet (t)

(b)

Multiplicity = $(N + 1)(M + 1) = 6$

Peak splitting pattern:
doublet of triplets (dt)

(c)

Multiplicity = $(N + 1)(M + 1)(P + 1) = 8$

Peak splitting pattern:
doublet of doublet of doublets (ddd)

FIGURE 7.19

Inverted tree diagrams for H_a in signals with various types of multiplicities, compared to (a) a standard triplet that follows the $N + 1$ rule. Relative intensities of each line within the multiplet are shown in red. When two nonequivalent sets of H's couple with H_a, the $N + 1$ rule is multiplied by $M + 1$, giving rise to more complex patterns. (b) When $N = 1$ and $M = 2$, a doublet of triplets (dt) arises. When a third different type of H is splitting, another multiplication with $P + 1$ is included. (c) When N, M, and P are all 1, the result is a multiplicity of 8, all with equal intensities, also known as a doublet of doublet of doublets (ddd).

RECIPROCITY OF COUPLING

An interesting feature of the spectrum of *n*-butyl vinyl ether (Figure 7.20) is that the same geminal coupling constant of 1.8 Hz is observed at both of the peaks at 4.2 and 3.9 ppm. This is a normal occurrence, because signal splitting between two H's is mutual, and the coupling constant J_{ab} observed at H_a will also be observed at H_b in the same magnitude. This feature can be useful when assigning a peak to a specific hydrogen in a structure. For example, knowing that H_b and H_c will have small geminal couplings allows us to assign them to the peaks at 4.2 and 3.9 ppm. However, which one is which? The third alkene peak at 6.4 ppm has coupling constants of 14.3 and 6.8 Hz that correspond to trans and cis relationships, respectively. The hydrogen that is trans to H_a should have the larger coupling constant 14.3 as seen at 4.2 ppm, so we assign this peak to H_b.

FIGURE 7.20

^{1}H NMR spectrum of *n*-butyl vinyl ether. H_a is a doublet of doublets because of coupling to cis (H_c) and trans (H_b) protons; H_c and H_b are nonequivalent and have different vicinal coupling constants with H_a. Expansion plots of the peaks at 4.2 and 3.9 ppm show evidence of geminal coupling ($J_{bc} = 1.8$ Hz).

© Sigma-Aldrich Co. LLC. Reproduced with permission from Merck KGaA, Darmstadt, Germany and/or its affiliates.

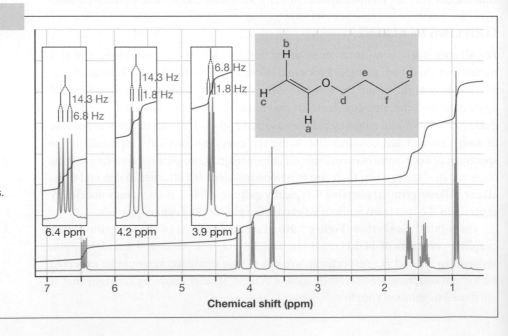

7.1G   Impact of Rapid C—C Rotation on Multiplicity

Based on the structure of *n*-butyl vinyl ether (Figure 7.20), what would you predict the multiplicities of H_e and H_f to be? H_e should be coupled to both H_d and H_f, and H_d and H_f are nonequivalent, so you might say that H_e should have a total multiplicity of

$$\text{Multiplicity} = (N + 1)(M + 1) = (2 + 1)(2 + 1) = 9$$

The single C—C bonds in the butyl group are rapidly rotating, however, which causes the coupling constants to be almost the same, at 7 Hz. This is because the coupling constant results from an average dihedral angle from a composite of all the rotamers. When the coupling constants are all indistinguishable, which is typical for *n*-alkyl groups, then the appearances of the multiplets can be better predicted by grouping N and M together. In this case, $N = 2$ (for H_d) and $M = 2$ (for H_f), and thus for H_e, the multiplicity is

$$\text{Multiplicity} = (N + M + 1) = (2 + 2 + 1) = 5$$

The peak appears as a quintet. Although it is not a true quintet, because the hydrogens represented by N and M are not equivalent, the appearance of the peak suggests that it seems to be following the $N + 1$ rule with $N = 4$. This is typical behavior for acyclic alkyl groups like the *n*-butyl group, and this greatly simplifies multiplicities, both in prediction and interpretation.

7.1H   Coupling and Chemical Exchange

Polarized bonds to hydrogen, such as N—H, O—H, and S—H, serve as hydrogen-bond donors, and can participate in *chemical exchange* of hydrogen as discussed previously in relation to analyzing the number of signals. If the exchange process is slow, it may not have any impact on vicinal coupling. In such circumstances, the O—H hydrogen in CH_3OH, for example, would be coupled to the methyl group, giving rise to a signal that is split into a quartet. However, the rate of chemical exchange can be faster at higher concentrations or in the presence of acid. When the exchange is rapid, the molecular environment around the hydrogen nucleus is rapidly changing, and the signal is formed from an average of all the various environments in which the hydrogen exists. This causes the signal to appear broadened, and also leads to the loss of coupling information. So, when interpreting spectra of such compounds, it's important to be aware that we don't necessarily know whether the spectrum was acquired under conditions that promote chemical exchange. Be prepared to see these signals broad or narrow, at variable chemical shift, and with or without signal splitting.

7.1I   Recap of ¹H NMR Data Interpretation

To summarize, there are four main types of information that can be obtained from the ¹H NMR spectrum:

1. Number of different kinds of hydrogens (number of signals).

2. Functional group and environments around these hydrogens (chemical shift).

3. Number of hydrogens in each environment (integration).

4. Number of hydrogens attached at the neighboring atom (signal splitting or multiplicity).

These are four very powerful pieces of information to help you determine the molecular structure of a compound. Given a formula of an unknown compound, which can be determined by other types of experimental techniques, NMR data can complete the picture in many cases, telling us how those atoms are connected and reveal the structure.

7.1J | Additional Real-Life Practical Situations with ¹H NMR Spectra

OVERLAPPING OR UNRESOLVED SIGNALS

The NMR spectra obtained from experimental samples are not always as perfectly behaved as those presented in a textbook. There may be hydrogens that are non-equivalent on the theoretical level, but in practice their environments and resonance energies are so similar that routine-use instruments cannot *resolve* the signals (i.e., cannot see separate peaks). In 1-heptanol (**Figure 7.21**), for example, the CH_2 units that are farther away from the functional group are not all resolved; instead, they produce a large peak of unclear multiplicity at 1.3 ppm that can be simply labeled "m" for multiplet. The integration, which is much greater than expected for a single CH_2 unit, is what indicates that several different signals are contained in this peak. In a simple structure like this, it may be unnecessary to resolve all of those CH_2 units. Spectra produced by higher field NMR instruments with more powerful magnets may be able to resolve all the signals of a very complicated structure, but for simpler compounds like 1-heptanol, the information gained may not be worth the expense of such an instrument.

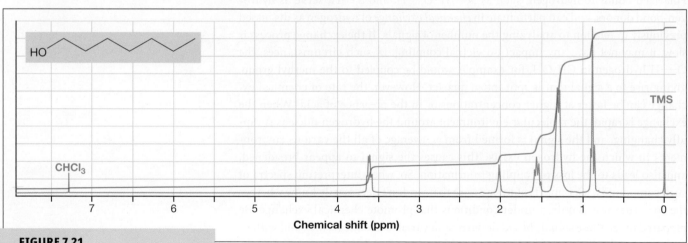

FIGURE 7.21

¹H NMR spectrum of 1-heptanol (300 MHz, CDCl₃). At first glance, it appears there are only five signals, and that seems inconsistent with the structure. However, the peak at 1.3 ppm actually contains four signals that are unresolved because their environments are very similar. The large integration at 1.3 ppm is the clue that should help alert you to situations like this.

© Sigma-Aldrich Co. LLC. Reproduced with permission from Merck KGaA, Darmstadt, Germany and/or its affiliates.

SOLVENT PEAKS AND CALIBRATION

Solvents used to prepare solutions for ¹H NMR spectroscopy are usually the same kinds of solvents normally found in a typical organic lab, except the hydrogens have been substituted with deuterium (²H, or D). Deuterium is the isotope of hydrogen that contains one proton and one neutron in the nucleus. The deuterium nucleus absorbs radiofrequency energy outside the range of the ¹H spectrum, so these solvents do not interfere with ¹H signals from the sample. Even with great care to replace all the hydrogens with deuterium, there will be traces of hydrogen left in the solvents, so a small peak from the solvent usually appears in the spectrum. The chemical shifts of these peaks are listed in **Table 7.1**, and they are often used to calibrate the frequency (*x* axis, in ppm) of the spectrum (notice the peak for CHCl₃ in Figure 7.21). Alternatively, a trace of (CH₃)₄Si (tetramethylsilane,

TMS) can be added to the solvent. TMS produces a singlet at 0.00 ppm (see Figure 7.21), out of the range of peaks observed in typical organic compounds, and this also serves as a standard for frequency calibration.

Trace amounts of water are generally seen in experimental ^1H NMR spectra. This may arise from incomplete drying of a sample before preparing the NMR tube, incomplete drying of the NMR tube after cleaning, or an impurity in the source bottle of the NMR solvent. Deuterated dimethylsulfoxide (DMSO-d_6) is notoriously hygroscopic, meaning that it absorbs moisture from humid air, so it often gives a very large water peak in the ^1H NMR spectrum. More nonpolar solvents such as $CDCl_3$ or benzene-d_6 (deuterated benzene) may be kept fairly dry, but even transferring the solvent to the NMR tube can lead to traces of moisture due to evaporative cooling and condensation at the pipet tip. As a result, you should always expect to see a water peak, and you should neglect it in the interpretation of the spectrum. To help you identify it, the chemical shifts for the water peak in various deuterated solvents are also listed in Table 7.1.

SECOND-ORDER NMR SPECTRA

Multiplet peak intensities in the sample spectra so far have been mostly very orderly, appearing in patterns that follow Pascal's triangle. As two sets of coupled hydrogens get closer in chemical shift, however, the individual peaks become distorted, so that it appears they are "leaning" toward each other; the inside edges of the set of two multiplets have a higher intensity. The series of spectra in **Figure 7.22** illustrates this effect for compounds with the general structure XCH_2CH_2Y, which would be expected to provide two triplets near 3 ppm. On the far left, the two triplets are about 0.6 ppm apart, and appear almost undistorted. From left to right, the chemical shifts of the two signals get closer together, leading to progressively more distorted multiplets. At the extreme case where X and Y are identical, there would be only one singlet because all four H's would be equivalent. This is an example of second-order effects, and the distortion caused by such effects can make it harder to determine multiplicity. On the other hand, the distortion is indicative of peaks that are coupled, so it can sometimes help in assigning which peaks belong to neighboring protons.

LONGER-RANGE COUPLING

In certain cases, coupling may be observed between hydrogens that are more than three bonds apart. The most common examples are seen with aromatic hydrogens, where small four-bond coupling constants can sometimes be observed in the range

TABLE 7.1

Common NMR Solvents and Their Peaks in ^1H NMR Spectra

DEUTERATED NMR SOLVENT	FORMULA	δ (ppm)	BOILING POINT (°C)
Acetone	$(CD_3)_2C{=}O$	2.05 2.84 (H_2O)	57
Benzene	C_6D_6	7.16 0.40 (H_2O)	80
Chloroform	$CDCl_3$	7.26 1.56 (H_2O)	61
Dimethyl sulfoxide	$(CD_3)_2S{=}O$	2.50 3.33 (H_2O)	189
Methanol	CD_3OD	3.31 4.87 (H_2O)	65
Water	D_2O	4.79	101

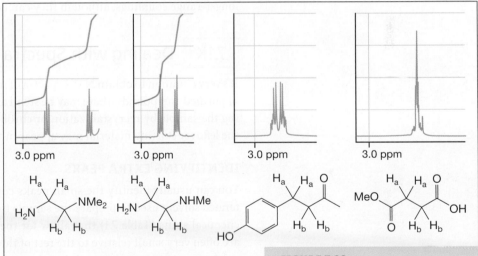

FIGURE 7.22

Distortion of multiplets in second order NMR spectra (300 MHz). From left to right, the chemical shifts of H_a and H_b become closer together, leading to a progressively more distorted appearance.

© Sigma-Aldrich Co. LLC. Reproduced with permission from Merck KGaA, Darmstadt, Germany and/or its affiliates.

FIGURE 7.23

(a) Hydrogens in aromatic rings and a comparison of their longer-range four-bond coupling constants with the standard vicinal 3-bond coupling of ortho hydrogens.
(b) Representative examples of doublet of doublet peaks arising from aromatic hydrogens with both 3- and 4-bond couplings.

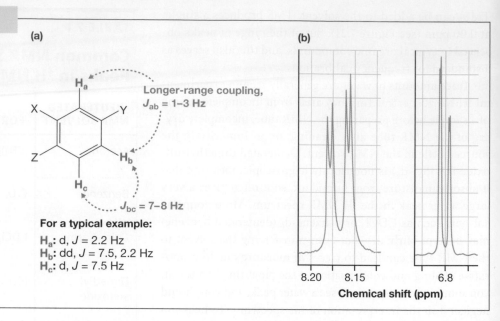

of $J = 1$–3 Hz (**Figure 7.23**). These small couplings are not always resolved in a spectrum. A peak that may look like a doublet with $J = 7$–8 Hz, when viewed more closely by "zooming in," may actually be a doublet of doublets with one very small coupling constant. Peak assignments can often be made without considering these longer-range couplings, although they can occasionally be useful.

7.1K   Dealing with Spectra of Mixture Samples

It's very common to obtain NMR spectra of mixtures. In experimental samples, even of purified compounds, there may be small amounts of the solvents used in processing the sample by recrystallization or chromatography. If a reaction is incomplete, the leftover starting material may appear in the NMR spectrum.

IDENTIFYING EXTRA PEAKS

You can usually identify the small peaks corresponding to residual ^1H in the deuterated solvent used to prepare the sample (see Figure 7.21). These have well-known chemical shifts (Table 7.1) that allow for these peaks to be assigned. Solvent peaks are often very small relative to the rest of the peaks in the spectrum, and their integration may appear to be much less than one H. It would be absurd to have a structure with one-tenth of a hydrogen in it, so when you have attempted to reduce all the integration data to integers, and a peak is left as a noninteger, this is a clue that it is not part of the same structure. It is likely an impurity. To identify it, gather NMR data (from literature or prediction) about all the components you used in the experiment: starting materials, reagents, solvents, etc. Do the same for any by-products expected. For example, if you are conducting an E1 elimination reaction, you may be trying to make an alkene but there could be products from the leaving group, or from S_N1 as a side reaction.

QUANTIFYING RATIOS IN MIXTURE SAMPLES

Integration helps to identify which peaks may be assigned to different components of a mixture. It also can be used to calculate the ratio of two components. The integration of two peaks within the same compound allows us to determine the mole ratio of hydrogens involved in those peaks. Within the same structure, all peak integrals should be equal on a per hydrogen basis. Similarly, for two peaks from two different compounds, the integration will be equal to the mole ratio of the hydrogens involved. First, the integrations are converted to a per hydrogen basis by dividing each by the number of equivalent hydrogens giving rise to the peak. Then, the ratio of these equals the mole ratio of the compounds in the mixture.

Example

In an esterification reaction to make ethyl acetate from ethanol and acetic acid, the ^1H NMR spectrum of the product (**Figure 7.24**) shows two peaks in the region expected for a CH_3CH_2—O group at 4.1 ppm and 3.7 ppm, suggesting the product is actually a mixture of ethyl acetate and ethanol. Calculate the ratios of the components.

To solve this problem, locate a peak from each of the components, and divide the integral values for those peaks by the number of H's assigned to those peaks. This gives you the integrations adjusted to a per hydrogen basis. The mole ratio is then given by the ratio of these two numbers. If desired, the mole ratio can be converted to a weight percentage by converting each of the mole amounts to masses by dividing each by the molecular weight. The calculations are shown in **Figure 7.25**.

Using this technique to measure ratios of components requires that the two components are both known. If their structures are unknown, you will be missing key information on (a) how many hydrogens give rise to the peak, and (b) molecular weight. These cases present much more advanced problems, requiring further separation and experimentation to identify the unknown components.

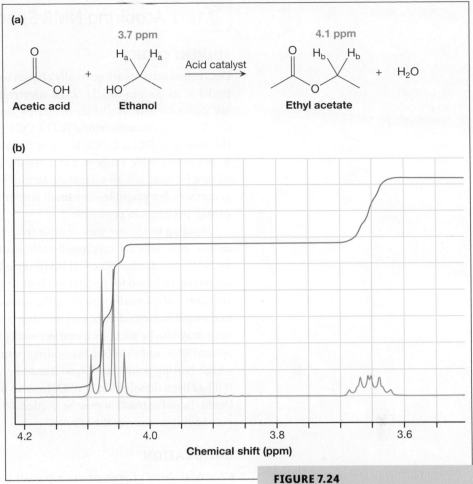

(a)   3.7 ppm   4.1 ppm

Acetic acid   Ethanol   Acid catalyst   Ethyl acetate   + H_2O

(b)

FIGURE 7.24

(a) The esterification reaction of acetic acid with ethanol. (b) A ^1H NMR spectrum showing a mixture of the product ethyl acetate and leftover ethanol. The integral values using arbitrary units are 260 units (4.07 ppm) and 120 units (3.65 ppm).

FIGURE 7.25

NMR integration data can be used to calculate weight percentage of components within a mixture.

$$\text{Mole ratio} = \frac{(\text{integral b})/\text{number of } H_b\text{'s}}{(\text{integral a})/\text{number of } H_a\text{'s}} = \frac{260}{120} = 2.17:1$$

This sample contains 2.17 moles of ethyl acetate for each mole of ethanol.

Because both integrals are measured on 2H peaks, the "number of H" terms cancel.

$$\text{Mass ratio} = \frac{2.17 \text{ mol EtOAc} \bullet 88 \text{ g/mol}}{1 \text{ mol EtOH} \bullet 46 \text{ g/mol}} = \frac{191 \text{ g}}{46 \text{ g}} = 4.15:1$$

This sample contains 4.15 g of ethyl acetate for each gram of ethanol.

$$\text{Weight \% ethyl acetate} = \frac{\text{mass EtOAc}}{\text{mass EtOAc} + \text{mass EtOH}} \times 100 = \frac{4.15 \text{ g}}{5.15 \text{ g}} \times 100 = 80.6\%$$

This sample is 80.6% ethyl acetate by weight.

7.1L | Acquiring NMR Spectra

SOLVENT CHOICE

Deuterated solvents are generally chosen because deuterium (^2H, or D) does not give peaks of its own in the ^1H NMR spectrum. A wide variety of deuterated solvents are commercially available, including deuterated dimethylsulfoxide (DMSO-d_6, CD_3SOCD_3), deuteroacetone (CD_3COCD_3), deuterobenzene (C_6D_6) and deuterochloroform ($CDCl_3$). $CDCl_3$ is one of the most widely used solvents in NMR because it dissolves a wide range of compounds. A small peak is always visible from the solvent because not all solvent molecules have had all the hydrogens replaced with deuterium. For example, very small amounts of $CHCl_3$ in the $CDCl_3$ will always give a small peak at 7.26 ppm.

Needing to recover your sample from the NMR tube may influence the solvent choice, especially for very small-scale work, where the entire sample may be used for NMR acquisition. Since ^1H NMR is nondestructive, the sample can generally be recovered and used for further analysis or reactions. In such cases, choose your solvent carefully! Usually removing the solvent is as simple as transferring the sample to a round-bottomed flask and swirling it while applying vacuum. A rotary evaporator may also be used. Evaporation works very well for solvents with boiling points around 80°C or below (see the boiling point data in Table 7.1). Note that DMSO has a high boiling point (189°C), so it may be very difficult to recover your compound if it has been dissolved in DMSO. Instead of using evaporation, a more complicated liquid–liquid extraction may be needed. When possible, avoid DMSO if you anticipate having to recover the sample.

CALIBRATION

Chemical shifts of protons are reported in parts per million (ppm) relative to the position of a sharp peak given by protons in tetramethylsilane (TMS), which is set at 0.00 ppm when the instrument is calibrated. Stock solutions of NMR solvents sometimes contain 0.05–3% TMS, so if a spectrum shows a peak at 0.00 ppm, it may be TMS. The spectrum can also be calibrated to the solvent peak; chemical shifts (in ppm from TMS) are known for all of the common solvents.

SAMPLE SPINNING

In order to average the magnetic fields produced by the spectrometer in the sample, the sample is spun inside the instrument at about 20 Hz, or 20 revolutions per second, while acquiring the spectrum. This produces a vortex in the tube which can interfere with acquisition, giving erratic, nonreproducible spectra. The effect of the vortex is minimized if the sample tube is filled to a minimum depth of about 2.5–3.0 cm, which keeps the vortex above the area where the radiofrequency receiver coil aligns with the sample tube (**Figure 7.26a**).

SAMPLE PREPARATION

After checking that a sample is soluble in the solvent of choice, a typical ^1H NMR sample is prepared by placing 5–20 mg of sample into a 5-mm-diameter tube. *CAUTION: NMR sample tubes are expensive and delicate!* The NMR solvent is then added to bring the level up to about 2.5–3.0 cm depth in the bottom of the tube (about 0.5–0.7 mL). If a sample contains some insoluble materials, solids floating around in the NMR tube will cause a very erratic spectrum. If necessary, solids should be removed by filtering the sample through a small plug of cotton in a Pasteur pipette. If you use the desired NMR solvent during this filtration, the filtration can be done directly into the NMR tube (**Figure 7.26b**).

 NMR Sample Preparation

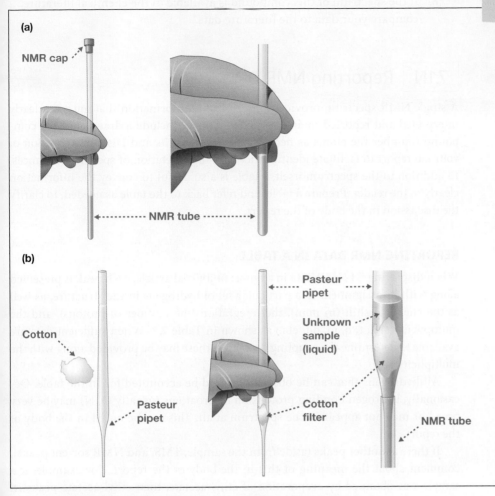

(a)

NMR cap

NMR tube

(b)

Cotton

Pasteur pipet

Pasteur pipet

Unknown sample (liquid)

Cotton filter

NMR tube

FIGURE 7.2G

(a) The sample solution should be a minimum of 2.5–3.0 cm depth in the NMR tube. (b) Filtering a solution into an NMR tube.

| 7.1M | **Interpretation of ¹H NMR Data: Characterization of a Known Compound** |

1. Draw the structure of the compound that you are trying to characterize. Count the total number of protons, and identify those that are equivalent.

2. Determine roughly what you would expect to see (number of peaks, chemical shift, integration, and multiplicity) for the structure you've drawn. Consider also any special circumstances, such as proton exchange or overlapping peaks, that could affect the appearance of the spectrum.

3. Identify solvent and calibration peaks (e.g., solvent, H_2O, TMS) in your spectrum and omit them from the analysis.

4. Convert peak integration data to an integer hydrogen count.

5. Determine the chemical shift of each peak in your spectrum.

6. Identify the multiplicity (splitting) of each resonance in the spectrum. You may also want to determine coupling constants (*J* values) for each set of resonances. This information can be used to determine which groups of protons are adjacent to each other in the structure.

7. Use all of this information to account for the resonances you see in the NMR spectrum. It may be helpful to consult a correlation table of chemical shifts.

8. If the spectrum of this compound is available in the chemical literature, compare your data to the literature data.

7.1N | Reporting NMR Data

A single NMR spectrum provides a great deal of information that must be clearly interpreted and reported in an organized manner. Include a drawing of the compound (number the atoms as necessary) in the Results and Discussion section of your lab report to facilitate identification and interpretation of specific resonances. In addition to the spectrum itself, a table is also useful to convey the information clearly to the reader. Prepare a table, and refer back to the table as needed, to clarify the discussion in the body of the report.

REPORTING NMR DATA IN A TABLE

When displaying ^{1}H NMR data in a report or journal article, each peak is presented along with its assignment to a particular set of hydrogens in the structure, as well as the chemical shift (in ppm), the integration (the number of protons), and the multiplicity (s, d, t, q, m, dd, etc.) as shown in **Table 7.2**. When sufficient data are available to determine the coupling constants, these may be provided along with the multiplicity.

All hydrogens that can be observed should be accounted for in the table. Occasionally, hydrogen bonding protons on heteroatoms (usually O, N) may be very broad or may not appear in the spectrum at all. This can be noted in the body of the report.

If there are other peaks (aside from the sample, TMS, and NMR solvent peaks), comment about the meaning of this in the body of the report. For example, if a sample was obtained by evaporation of ethyl acetate, then ethyl acetate peaks in the spectrum would indicate that the sample was not completely free of the solvent. While it may not be possible to assign every set of protons to a specific resonance (they may be grouped together, overlapping, or difficult to distinguish), every proton should be accounted for in the integration. The integration must be reported in whole numbers, and it should agree with the number present in the compound being evaluated. If these values do not agree, you must explain why.

TABLE 7.2

Peak Assignments and ^{1}H NMR Data for *n*-Butyl Vinyl Ether (See Figure 7.20)

CHEMICAL SHIFT	INTEGRATION	MULTIPLICITY	ASSIGNMENT
0.9 ppm	3	t (J_{fg} = 7 Hz)	H_g
1.4 ppm	2	m (apparent sextet, J = 7 Hz)	H_f
1.6 ppm	2	m (apparent quintet, J = 7 Hz)	H_e
3.7 ppm	2	t (J_{de} = 7 Hz)	H_d
3.9 ppm	1	dd (J_{ac} = 6.8 Hz, J_{bc} = 1.8 Hz)	H_c
4.2 ppm	1	dd (J_{ab} = 14.3 Hz, J_{bc} = 1.8 Hz)	H_b
6.4 ppm	1	dd (J_{ab} = 14.3 Hz, J_{ac} = 6.8 Hz)	H_a

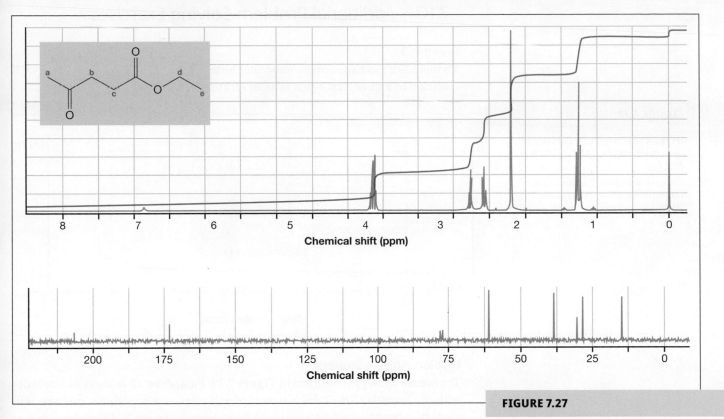

FIGURE 7.27

^{1}H and ^{13}C NMR spectra of ethyl levulinate. The ^{13}C NMR spectrum (bottom) has a solvent peak (CDCl$_3$) appearing as three peaks centered at 77 ppm.

© Sigma-Aldrich Co. LLC. Reproduced with permission from Merck KGaA, Darmstadt, Germany and/or its affiliates.

Example

The ^{1}H and ^{13}C NMR spectra (300 MHz, CDCl$_3$) of ethyl levulinate are presented in **Figure 7.27**. **Table 7.3** and **Table 7.4** show how these NMR data should be tabulated in your report.

TABLE 7.3

^{1}H NMR Data for Ethyl Levulinate

CHEMICAL SHIFT	INTEGRATION	MULTIPLICITY	ASSIGNMENT
1.25 ppm	3H	t	—CH$_3$ (e)
2.19 ppm	3H	s	—(C=O)CH$_3$ (a)
2.55 ppm	2H	t	—CH$_2$(C=O) (b or c)
2.74 ppm	2H	t	—CH$_2$(C=O) (b or c)
4.15 ppm	2H	q	—OCH$_2$ (d)

TABLE 7.4

^{13}C NMR Data for Ethyl Levulinate

CHEMICAL SHIFT	ASSIGNMENT
14.2 ppm	—CH$_3$ (e)
28.1 ppm	—CH$_2$(C=O) (c)
29.8 ppm	—(C=O)CH$_3$ (a)
38.0 ppm	—CH$_2$(C=O) (b)
60.5 ppm	—OCH$_2$ (d)
172.7 ppm	C=O, ester
206.6 ppm	C=O, ketone

Example 3

An unknown aldehyde has the molecular formula C_4H_8O and exhibits the ^{1}H NMR spectrum in **Figure 7.28**. What is its structure?

FIGURE 7.28

^{1}H NMR spectrum of an aldehyde with the formula C_4H_8O (300 MHz, CDCl$_3$).

© Sigma-Aldrich Co. LLC. Reproduced with permission from Merck KGaA, Darmstadt, Germany and/or its affiliates.

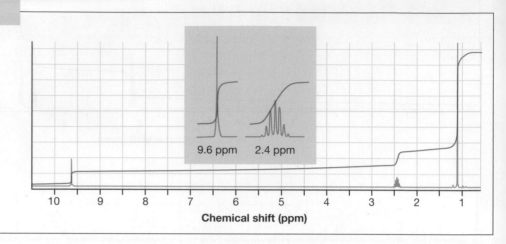

Problem-Solving Strategy

The overall strategy is outlined in **Figure 7.29**. First, draw all isomers of aldehydes with the formula C_4H_8O. Each structure will serve as a hypothesis, and you will apply the scientific method to test each hypothesis, using the NMR evidence to systematically disprove all but one. To do this, predict the number of signals, chemical shift, integration, and multiplicity for each structure. Then, compare these predictions with the observed spectrum. If there is any inconsistency between prediction

FIGURE 7.29

Problem-solving strategy for identifying an unknown. Here the unknown is an aldehyde of formula C_4H_8O having the spectrum shown in Figure 7.28. In this case only the number of signals was needed to identify the structure by process of elimination.

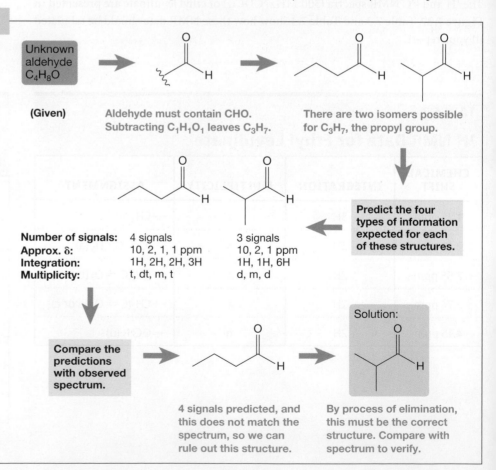

and observation, that structure can be ruled out (disproven). When you are left with only one structure that cannot be ruled out, you have solved the problem. Disproving a hypothesis requires only one inconsistency between prediction and observation. Thus, you may not need to use all four types of information that the NMR spectrum provides.

Example 4
An unknown alcohol has the molecular formula $C_8H_{10}O$ and exhibits the ^1H NMR spectrum in **Figure 7.30**. What is its structure?

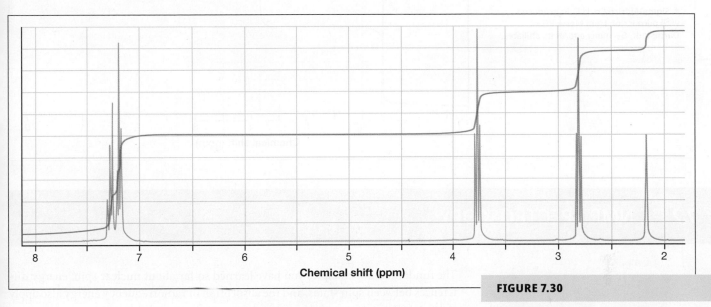

Chemical shift (ppm)

FIGURE 7.30

^1H NMR spectrum of a compound of an alcohol with the formula $C_8H_{10}O$ (300 MHz, CDCl$_3$).

Example 5
An unknown ester has the molecular formula $C_5H_8O_2$ and exhibits the ^1H NMR spectrum in **Figure 7.31**. What is its structure?

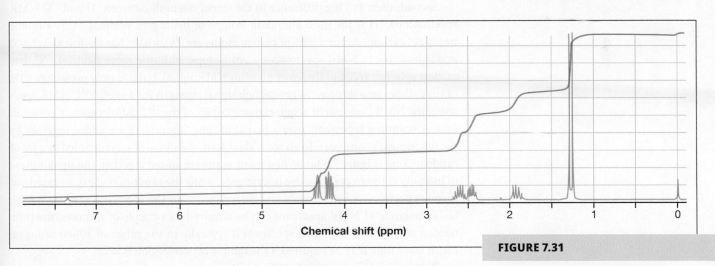

Chemical shift (ppm)

FIGURE 7.31

^1H NMR spectrum of an ester with the formula $C_5H_8O_2$ (300 MHz, CDCl$_3$).

Example 6

An unknown compound, insoluble in water and soluble in 10% aqueous HCl, has the molecular formula $C_6H_8N_2$ and exhibits the ^1H NMR spectrum in **Figure 7.32**. What is its structure?

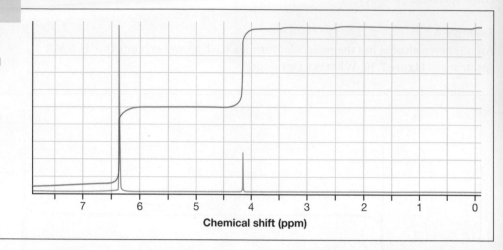

7.2 ^{13}C NMR SPECTROSCOPY

The fundamental principles you have learned so far about nuclear spin, energy differences between spin states, and the absorption of radiofrequency energy also apply to ^{13}C NMR spectroscopy. However, there are some key differences.

First, absorption of energy by ^1H and ^{13}C nuclei in a magnetic field occurs in different areas of the radiofrequency range, so ^1H and ^{13}C peaks do not appear on the same spectrum. Compared to ^1H NMR, the peaks in ^{13}C NMR appear over a much wider δ range, generally from –10 ppm to 220 ppm.

Second, there is a big difference in the signal strength between ^1H and ^{13}C NMR spectroscopy. ^1H is the most abundant isotope of hydrogen, whereas ^{13}C is a minor isotope of carbon. Around 99% of carbon atoms are ^{12}C, which has nuclear spin $I = 0$ and is invisible to NMR spectroscopy. An isotope with one more neutron, ^{13}C, has nuclear spin $I = 1/2$, and therefore gives an NMR signal, but it is only present as 1.1% of the carbons in a sample. To get enough signal strength for a useful ^{13}C NMR spectrum, the NMR instrument acquires many copies of the ^{13}C radiofrequency emission data by scanning repeatedly over a period of time. These data are added together by the instrument's computer. In an individual scan, the signal is accompanied by a lot of random noise (**Figure 7.33a**). When many scans are added together, the signals grow in intensity through constructive interference, while the random noise is averaged out through destructive interference and disappears from the spectrum (**Figure 7.33b**). So, although a ^1H NMR spectrum can be acquired in a couple of minutes, the time needed to acquire a ^{13}C NMR spectrum is typically in the range of 30–60 minutes. Much more time may be required if the sample concentration is low.

Third, ^1H NMR differs from ^{13}C NMR in coupling and multiplicity. Among the 1.1% of carbons that are ^{13}C, there is only a 1.1% chance of having another ^{13}C nucleus as its neighbor. Thus, there is no significant coupling between carbons. When a ^{13}C nucleus is attached to another carbon, it is almost always the more abundant ^{12}C isotope that has no nuclear spin. On the other hand, nearby ^1H nuclei may be coupled with ^{13}C, and the coupling constants are large enough that the signal splittings can cause the peaks to overlap throughout the spectrum. This makes it very difficult to

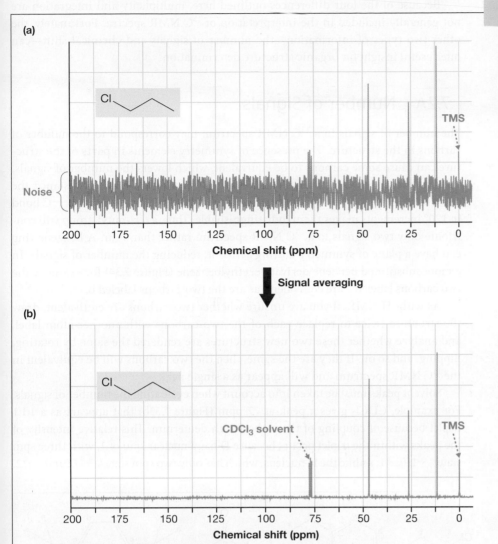

FIGURE 7.33

The ^{13}C NMR spectrum of 1-chloropropane. Peaks at 0 ppm and 77 ppm are from the solvent and the calibration standard.

Karty, J. *Organic Chemistry: Principles and Mechanisms*, 3rd ed.; W. W. Norton: New York, 2022; p 854.

get useful information from the spectrum. To avoid this, ^{13}C NMR spectra are usually acquired with the ^{1}H and ^{13}C nuclei decoupled. This is accomplished by applying radiofrequency irradiation in the frequency range of the ^{1}H nuclei while simultaneously acquiring data in the ^{13}C range. All of the peaks in the resulting ^{13}C NMR spectrum appear as sharp singlets. Another benefit of decoupling is the improved signal-to-noise ratio when the peaks are condensed into singlets.

A fourth difference has to do with the intensity of ^{13}C signals within a spectrum of a given compound. Within a molecule, some nuclei relax from the excited β state to the lower energy α state more rapidly than others. If a spectrum was based on only one scan, this wouldn't be a problem. However, if another scan takes place before one type of nucleus has relaxed to its original α and β population distribution, then that nucleus gives less of a response in the instrument in the next scan, and its signal intensity will be weaker. In a spectrum that requires many scans, the differences are magnified in the ^{13}C NMR spectrum, and signal intensity from one carbon to the next can be vastly different. The signal intensities are related to how many hydrogens are attached at the carbons. Those carbons with no hydrogens attached, such as the 3° carbon in *tert*-butyl alcohol or the carbonyl carbon in a ketone, tend to have greatly diminished signal intensity relative to a CH_2 or CH_3. This can be helpful in assigning peaks in the ^{13}C NMR spectrum. Quantifying these intensity differences is generally not useful, so integration is not commonly employed in ^{13}C NMR.

Because of the four differences outlined here, multiplicity and integration are not generally included in the interpretation of ^{13}C NMR spectra. Fortunately, the other two types of information—the number of signals and chemical shift—can offer useful insight for organic structure determination.

7.2A Number of Signals

The number of signals in a ^{13}C NMR spectrum may correspond to the number of carbons in the structure. The presence of symmetry elements in parts of the structure can cause some carbons to be equivalent, which lowers the number of signals, just as it does in ^{1}H NMR. Some examples are given in **Figure 7.34**. For instance, the three methyls of a *tert*-butyl group are equivalent because rotation of a C—C bond in 120° increments makes them indistinguishable. Thus, a *tert*-butyl group will contribute only two signals to a ^{13}C NMR spectrum rather than four. A benzene ring can have a plane of symmetry cutting across it, reducing the number of signals. In a monosubstituted benzene derivative, ethylbenzene (Figure 7.34) for example, the two carbons labeled c are equivalent, as are the two carbons labeled b.

As with ^{1}H NMR, if you are unsure whether two carbons are equivalent, draw the structure twice to replace each of the two carbons with another atom label, and analyze whether these two new structures are rendered the same by rotating, flipping, and so on. If they are the same, then the two carbons will be equivalent in the ^{13}C NMR spectrum and will appear as a single peak.

Solvent peaks must be taken into account when evaluating the number of signals. For example, CDCl$_3$ gives a peak at 77 ppm (Figure 7.33b) that appears as a 1:1:1 triplet because of coupling of the carbon with deuterium. This relative intensity of the peaks within the triplet occurs because D has a nuclear spin of 1, with three spin states, −1, 0, +1, unlike the H nucleus, which has only two spin states, −1/2 and +1/2.

FIGURE 7.34

Examples illustrating the number of nonequivalent carbons in a variety of structures. Nonequivalent carbons are labeled in red letters. Blue letters indicate carbons that are equivalent to those with the same letter labeled in red.

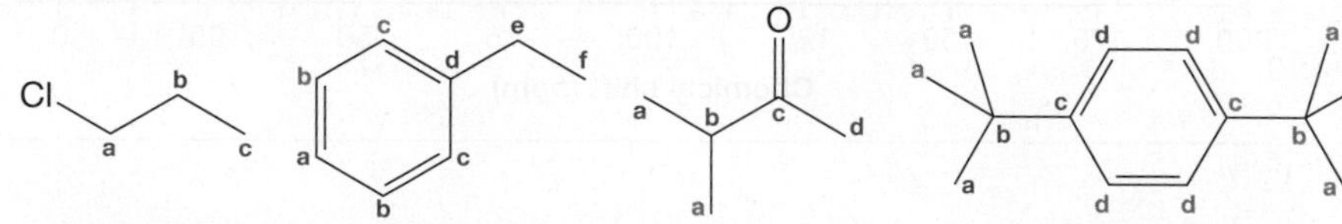

1-Chloropropane   **Ethylbenzene**   **3-Methylbutanone**   ***p*-Di(*tert*-butyl)benzene**

7.2B Chemical Shift

Chemical shift in ^{13}C NMR correlates to shielding and deshielding, as in ^{1}H NMR, and rough estimates of chemical shifts of various functional groups are summarized in **Figure 7.35**. The effects of electronegativity on deshielding can be observed in the comparisons of chemical shifts of alcohols (60–80 ppm) to amines (30–60 ppm). Oxygen is more electronegative than nitrogen, and oxygen shifts the carbon absorbance farther downfield. These deshielding effects are additive, as in ^{1}H NMR, so two oxygens combine to shift acetal carbons even farther downfield (80–100 ppm). Hybridization and magnetic anisotropy combine with electronegativity to impact the chemical shifts of π systems such as alkenes, aromatic rings, and carbonyl compounds.

Certain areas of the ^{13}C NMR spectrum are exceptionally useful in diagnosing functional groups. For example, the carbonyl carbon of aldehydes and ketones (190–220 ppm) is readily differentiated from that of carboxylic acids and their

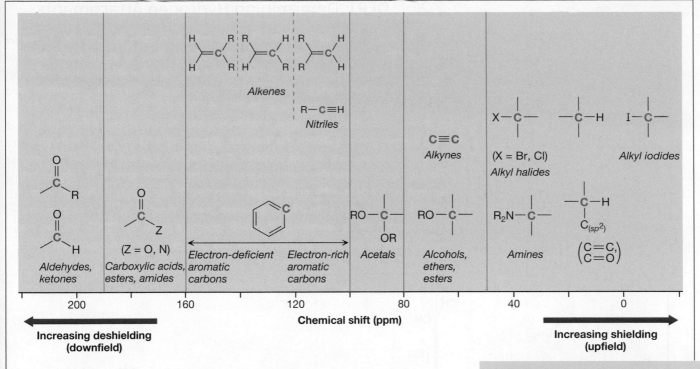

FIGURE 7.35

Typical ^{13}C NMR chemical shift ranges for some common functional groups.

derivatives (160–180 ppm, includes esters, amides, acyl halides, and anhydrides), complementing the information available from ^{1}H NMR, which does not directly display a peak for the C=O.

Drawing resonance structures can be very helpful in assigning signals to carbons within conjugated π systems such as unsaturated carbonyl compounds. For 2-cyclohexen-1-one (**Figure 7.36**), carbons at positions 2 and 3, though both alkene carbons, have a dramatic 21 ppm difference in their chemical shifts. The partial charges in the resonance hybrid reveal why this is so. There is greater electron density at C-2, and more shielding, because of its partial negative charge; C-2 appears at 129.8 ppm. On the other hand, C-3 has a partial positive charge, and this electron-deficient character causes it to be deshielded in comparison with C-2; C-3 appears at 150.9 ppm. Similar effects can be seen in a variety of substituted π systems with significant contributions of charged resonance structures, including benzene derivatives.

FIGURE 7.36

Effects of charge distribution on the chemical shift in the π system of 2-cyclohexen-1-one.

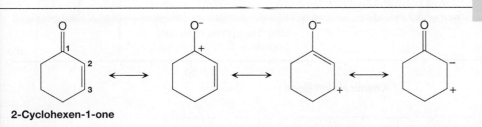

2-Cyclohexen-1-one

Carbon	δ (ppm)
C-1	199.7
C-2	129.8
C-3	150.9

Resonance hybrid

7.2C | DEPT: Determining Hydrogen Attachments

All of the peaks in a typical ^{13}C NMR spectrum are singlets because the ^{13}C nuclei are decoupled from the ^{1}H nuclei. It can be difficult, then, to assign some peaks to specific carbons in the structure. A helpful tool for this job is *distortionless enhancement by polarization transfer* (DEPT) NMR spectroscopy. The most commonly used format of DEPT NMR is known as DEPT-135. This is a variant of the ^{13}C NMR experiment, and carbons show up as singlets as usual, but are phased differently so that peaks appear above or below the baseline depending on how many hydrogens are attached. Carbons with an odd number of hydrogens attached (CH or CH$_3$) appear above the baseline, and those with an even number of hydrogens (CH$_2$) appear below the baseline. Carbons with no hydrogens attached do not appear in the spectrum. All of the chemical shifts in the DEPT spectrum match those of the ^{13}C NMR spectrum. So, by comparing the standard ^{13}C NMR spectrum with the DEPT-135 spectrum, you can identify the types of substitution that are associated with each peak.

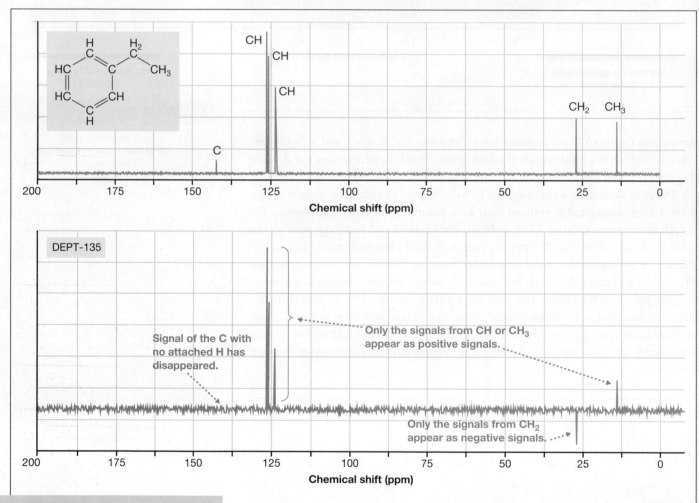

FIGURE 7.37

^{13}C NMR and DEPT-135 NMR spectra of ethylbenzene.

Karty, J. *Organic Chemistry: Principles and Mechanisms*, 2nd ed.; W. W. Norton: New York, 2018; p 810. Karty, J. *Organic Chemistry: Principles and Mechanisms*, 3rd ed.; W. W. Norton: New York, 2022; p 858.

For example, the ^{13}C NMR spectrum of ethylbenzene (**Figure 7.37**) has four signals in the 120–140 ppm range and two in the 10–30 ppm range. Based on the chemical shifts, these can be assigned to the aromatic carbons and the ethyl group, respectively. But how can we determine which of the ethyl group signals belongs to the CH$_2$? In the DEPT-135 spectrum, there is only one peak below the baseline, so that peak corresponds to the CH$_2$ group. Moreover, the ring carbon at the point of attachment of the ethyl group has disappeared from the DEPT-135 spectrum, because it has no hydrogens attached.

8

125

LEARNING OBJECTIVES

- Describe how mass spectrometry can be used to determine a molecular formula.

- Predict fragmentation processes associated with structural features of organic compounds.

- Distinguish among possible molecular formulas, using high-resolution mass spectrometry.

- Use combustion analysis data to determine molecular formulas.

- Evaluate purity of an organic compound, using combustion analysis.

- Identify degrees of unsaturation from a molecular formula and apply it to structure determination.

- Use molecular mass and formula for structure determination in combination with spectroscopy.

Determination of Molecular Mass and Formula

MASS SPECTROMETRY INSTRUMENTS

Mass spectrometers are traditionally cost- and space-intensive tools for identifying organic compounds, but newer instruments continue to become more widely accessible.

RICHARD NOWITZ/Science Source.

I n the identification of an organic compound's structure, a critical piece of information is its molecular formula, or the ratio of the elements present in the compound. Two example formulas are C_6H_{14} for hexane and $C_{21}H_{22}N_2O_2$ for strychnine (**Figure 8.1**), a natural alkaloid known for its pesticide properties.

Strychnine
$C_{21}H_{22}N_2O_2$

Diethyl ether
$C_4H_{10}O$

Triethylamine
$C_6H_{15}N$

What can we do with this molecular formula? Given some sample history, such as the source of the compound and the conditions of its reactions or processing, there may be a short list of possible compound structures, and determining the molecular formula can often distinguish between those possibilities. It should be noted that isomers will have the same molecular formula, and cannot be distinguished in this manner. For example, ethanol and dimethyl ether each have the formula C_2H_6O, but have different connectivity of the atoms, and the enantiomers (R)-2-butanol and (S)-2-butanol are different configurations of the same formula and connectivity. Neither of these pairs can be distinguished by molecular formula. Still, there are many situations where the molecular formula will provide a key to identifying the compound in question.

Without any sample history, some additional data from spectroscopy (for example, IR and NMR) would generally need to accompany the formula in order to provide enough information for a confident structural assignment. In some cases, the formula is the deciding factor. Consider two colorless liquids that exhibit IR and NMR spectra that are quite similar: diethyl ether and triethylamine (Figure 8.1). Outside of the fingerprint region, neither has obvious **diagnostic peaks** in the IR spectrum. In their ^1H NMR spectra, both would exhibit the readily recognized pattern of an ethyl group attached to an electronegative atom—an upfield triplet and a downfield quartet. But the molecular formulas of these two compounds are very different; one contains nitrogen and the other oxygen. Determining the molecular formula experimentally would easily distinguish between these two compounds.

In this chapter we will examine two methods to determine a molecular formula experimentally. First, we will see how ions may be generated from organic molecules in the gas phase, how the masses of these ions are measured via mass spectrometry, and how the data are correlated to molecular formula. Second, we will learn about a technique known as combustion analysis, in which the organic compound is burned under very controlled conditions that allow the measurement of the percent by weight of carbon, hydrogen, and nitrogen in the sample. This too can be correlated to molecular formula. With the molecular formula in hand, the chemist is armed with the information needed to either confirm a structure or propose a short list of possible structures for further evaluation.

diagnostic peaks >>
In spectroscopy, peaks that are characteristic of a compound and can be used for its identification or quantification. In comparing reactants and products, the diagnostic peaks are usually those that are closely associated with the functional group where the reaction occurred.

8.1 MASS SPECTROMETRY

8.1A Introduction: Separation of Ions by Mass

Organic compounds can be converted into positively or negatively charged ions in the gas phase. When these charged particles are generated in a vacuum chamber, their travel through the chamber can be accelerated and directed by electrostatic and magnetic fields, causing the ions to pass through the chamber toward a detector. The **relative abundance** of different ions that reach the detector, sorted by mass, can be recorded in a spectrum. The ion current or quantity of ions detected is plotted on the y axis versus the mass-to-charge ratio (m/z) of the ions on the x axis. Although an ion may have a net charge of two or more, either positive or negative, in typical organic chemistry applications we will consider particles that have a charge $z = +1$. In this case the mass-to-charge ratio m/z simplifies to just mass. Although we will refer to mass of ions in this chapter, it is good to be aware that *the mass spectrometer is actually measuring m/z*.

At the core of the mass spectrometry instrument is an analyzer device that separates ions as a function of m/z, so that ions of differing mass are detected separately and their relative abundance can be recorded in a **mass spectrum** (**Figure 8.2b**). There are several different types of analyzers, including magnetic sector, quadrupole, ion trap, and time-of-flight (TOF). As this chapter is focused on interpretation of mass spectra for structure determination, we will not include details on the physics

<< relative abundance
In mass spectrometry, the ratio of the intensity of a smaller peak to the intensity of the largest peak, usually expressed as a percentage.

<< mass spectrum
A plot of masses of molecular ions (or more precisely, their mass-to-charge ratio m/z) versus their relative abundance. Such a plot is useful in identification and quantification of organic compounds.

FIGURE 8.2

(a) A mass spectrometer, circa 1989, and (b) an example of a mass spectrum.

Gado Images/Alamy Stock Photo. Karty, J. *Organic Chemistry: Principles and Mechanisms*, 3rd ed.; W. W. Norton: New York, 2022; p 742.

(a)

(b)

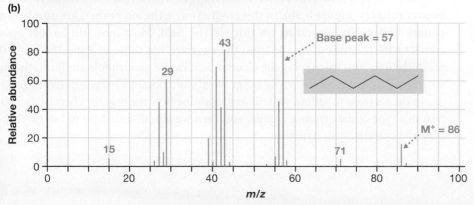

behind all of these analyzers. However, some explanation of magnetic sector and quadrupole analyzers will aid in understanding the principles of mass spectrometry.

The *magnetic sector analyzer* (**Figure 8.3a**) uses electrostatic fields to focus and accelerate ions into a specific flight path, or ion beam. A magnetic field then deflects the ion beam, generating a semicircular flight path that passes through a curved tube leading to a detector that measures the ion current. This deflection is related to the m/z and velocity of the particles as well as the strength of the applied magnetic field. If those variables yield an ion beam that matches the curvature of the tube, the ions reach the detector, where an ion current is registered. By systematically varying electrostatic or magnetic fields, the instrument can scan the ion current across a range of m/z. The instrument then processes the signal into a plot of ion current versus m/z to yield the mass spectrum.

FIGURE 8.3

(a) Schematic of electron impact mass spectrometer with magnetic sector analyzer, and (b) a quadrupole analyzer, in which the parallel charged rods cause ions to travel in helical paths. In both cases, the paths vary so that ions of specific m/z are detected separately.

Karty, J. *Organic Chemistry: Principles and Mechanisms*, 3rd ed.; W. W. Norton: New York, 2022; p 740. Taouatas, N.; Drugan, M. M.; Heck, A. J. R.; Mohammed, S. Straightforward Ladder Sequencing of Peptides Using a Lys-N Metalloendopeptidase. *Nat Methods* © **2008**, *5*, 405–407. https://doi.org /10.1038/nmeth.1204, reprinted by permission from Springer Nature.

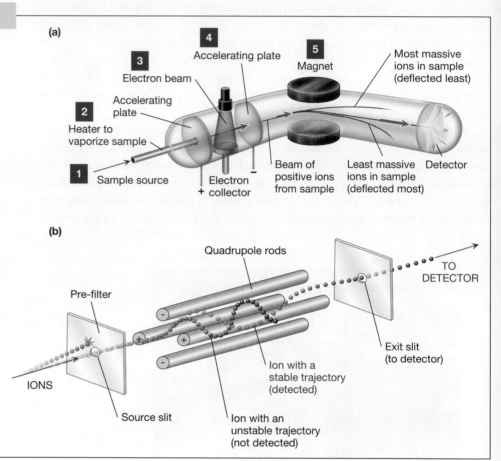

A *quadrupole analyzer* passes the ion beam through an electric field that is created between a set of four parallel rods charged with direct current and radiofrequency voltages (**Figure 8.3b**). In this environment, the ion beam takes on a helical pattern that varies depending on this electric field. Most ions will be lost upon contact with the rods or escape from the analyzer, but those of a specific m/z will reach the detector. By varying the electric field in the quadrupole analyzer, the instrument can detect ions across a range of m/z and obtain a mass spectrum.

In the sections to follow, we will first examine how the organic chemist can obtain this type of data from a sample, and then we will discuss how to interpret the data and correlate it to structure.

8.1B | Ionization Techniques

There are many different ways to generate ions from a sample of organic material. In early mass spectrometers, ions were generated by electron ionization (EI), also known as electron impact, by subjecting the sample to an electron beam. In collisions with the organic molecules, the high-energy electrons can knock an electron out of the bonding orbitals within the organic molecule, forming an ion known as a molecular ion (M^+):

$$M + e^- \rightarrow M^+ + 2\,e^-$$

The high energies involved usually cause the ions to fragment into many smaller ions and radicals; the ones that retain a charge continue on through the instrument. The masses that are detected are those of the fragments derived from the molecular ion. For an unknown compound, the mental exercise of putting these fragments back together can enable the chemist to derive structural information.

To better preserve the intact molecular ion, a softer ionization technique known as chemical ionization (CI) can be advantageous. Here, electron ionization takes place first on a reagent gas that then causes ionization of the sample. With methane as the reagent gas, CH_4^+ is formed, and this reacts with another methane molecule, transferring a hydrogen atom to give CH_5^+. This in turn transfers a proton to an analyte molecule, leading to an ion of formula $[M+H]^+$:

$$M + CH_5^+ \rightarrow [M+H]^+ + CH_4$$

Although this ion contains an extra hydrogen and is one mass unit more than M^+, it is still called the molecular ion because no fragmentation has occurred. Numerous reagent gases, in addition to methane, can be used for CI, including isobutane, ammonia, water, and others.

Electrospray ionization (ESI) is a method common to many newer instruments that does an outstanding job of retaining the molecular ion. Here, a solution of the sample passes through a metal capillary, causing charged **aerosol** particles to form. As these travel into the vacuum chamber, the solvent is removed, leaving behind charged particles of the sample that fly through the analyzer. As these are generated from solutions containing aqueous sodium ions, the molecular ions generally contain an extra proton ($[M+H]^+$) or sodium ion ($[M+Na]^+$). A mass spectrum using ESI often produces only molecular ions, so it is very useful for more complex samples that would give hopelessly complicated spectra if fragmentation occurred. As such, it is very useful for the analysis of mixtures, including biological analytes containing large proteins and other sensitive biopolymers. This type of instrument is often directly interfaced with high-performance liquid chromatography (HPLC); as the components of a mixture emerge from the chromatographic separation, they are detected by the mass spectrometer.

Finally, matrix-assisted laser desorption ionization (MALDI) is a method that permits analysis of very large ions up to m/z 300,000. The sample is placed within a matrix of crystalline material, often a carboxylic acid, that can absorb UV light. A UV pulse from a laser then causes rapid heating of the matrix and sample, forming gas-phase ions in a controlled fashion with little fragmentation. Since the materials used to form the matrix obscure the region from m/z 100 to m/z 500, this technique is more suited to polymers including biopolymers such as DNA, proteins, etc., and less often used for smaller molecules typically encountered in the organic chemistry lab.

<< aerosol
Finely divided liquid droplets, as in a spray or mist.

8.1C Practical Aspects of Sampling and Instrumentation

One of the great advantages of mass spectrometry is that vanishingly small samples can be analyzed effectively. A spectrum can be obtained on less than 0.1 mg of sample, a mass too small to accurately measure with a typical analytical balance. This is fortunate, because it is a destructive technique—sample recovery is generally not possible.

Sample preparation depends on the type of mass spectrometer you will use. In some larger laboratories or universities, mass spectrometers are operated by dedicated staff, and you may simply place a few milligrams of a sample in a labeled vial and submit it for analysis. If you'll be using the instrument yourself, ask your instructor what type of sample preparation will be needed. Some mass spectrometers allow for solid or liquid samples to be placed on a probe that is inserted directly into the mass spectrometer. Many modern instruments connect mass spectrometry with separations tools such as gas chromatography (GC-MS) or high-performance liquid chromatography (HPLC-MS), so that the eluent from the separation is directly passed into the mass spectrometer, affording a molecular formula of each separated component of a mixture. For these instruments, simply prepare the sample as you normally would for the separation. For GC, this generally involves dissolving the sample in a volatile solvent such as pentane or diethyl ether. For HPLC, check the type of column and its recommended mobile phase solvent, and prepare a solution using that solvent. Concentrations in the range of 1–10 mg/mL are normal, and only a few microliters of this solution will typically be used.

8.1D The Mass Spectrum

What information do most organic chemists seek in the mass spectrum? In most cases, the molecular ion is generally the primary source of information in the spectrum because it allows a determination of the molecular weight of the compound—information that is not available from IR or NMR spectra. Several other types of information can be accessed from routine mass spectra. These include extra peaks of higher m/z that indicate the presence of certain elements having heavier isotope(s) in natural abundance, such as chlorine, bromine, or sulfur. Mass spectra are also used to detect and measure heavier isotopes used to synthetically label compounds for studies of reaction mechanisms or biosynthetic pathways, such as ^{2}H (deuterium, D) or ^{13}C. Third, the fragmentation of organic ions during mass spectrometry can allow for identification of connectivity in small organic molecules, and thus distinguish between isomers. Fragmentation can also reveal the sequence of amino acids or nucleic acids in biopolymers such as proteins and DNA.

THE MOLECULAR ION

In an example of an electron impact mass spectrum (**Figure 8.4**), peak intensity on the y axis is normalized to the highest peak, which is assigned a relative abundance of 100%, and is called the *base peak*. Other peaks in the spectrum have intensities expressed as a percentage in relation to the base peak. Depending on how much fragmentation is occurring under the given ionization conditions, the base peak may or may not be the molecular ion. For a pure substance, usually the molecular ion is found in a cluster of peaks located at the highest mass in a mass spectrum, and corresponds to the species present after ionization of a molecule M but before any fragmentation occurs. Depending on the ionization method, this can take several forms, most commonly M^+, $[M+H]^+$, or $[M+Na]^+$ (see the section above on ionization techniques).

FIGURE 8.4

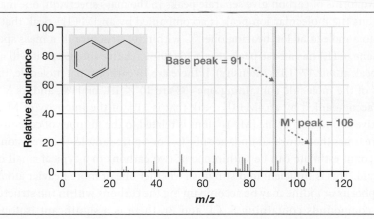

An electron impact ionization (EI) mass spectrum of ethylbenzene. The base peak at *m/z* 91 corresponds to $C_7H_7^+$, and the M^+ peak is observed at *m/z* 106 (relative abundance 28% of the base peak).

Karty, J. *Organic Chemistry: Principles and Mechanisms*, 3rd ed.; W. W. Norton: New York, 2022; p 744.

ISOTOPE PEAKS

The molecular ion peak is accompanied by several smaller peaks of masses one or two units higher than the main molecular ion peak; these are isotope peaks. Different isotopes result from extra neutron(s) in the nucleus of an atom. Some elements have significant amounts of more than one isotope. In typical samples of carbon found in nature, 98.9% of the carbon atom is ^{12}C, with atomic mass of 12 g/mol, while 1.1% of the carbon atoms is ^{13}C, one mass unit higher because it carries an extra neutron in its nucleus (**Table 8.1**). Thus we say the natural abundance of the ^{13}C isotope is 1.1%.

TABLE 8.1

Isotope Exact Masses and Their Relative Abundance

ELEMENT	ATOMIC MASS	ISOTOPE	ISOTOPE EXACT MASS	RELATIVE ABUNDANCE
Hydrogen (deuterium)	1.00797	^1H ^2H (D)	1.00783 2.0141	100 0.015
Carbon	12.01115	^{12}C ^{13}C	12.00000 13.0034	100 1.11
Nitrogen	14.007	^{14}N ^{15}N	14.0031 15.0001	100 0.37
Oxygen	15.999	^{16}O ^{18}O	15.9949 17.9992	100 0.20
Sulfur	32.064	^{32}S ^{33}S ^{34}S	31.9721 32.9715 33.9679	100 0.79 4.43
Chlorine	35.453	^{35}Cl ^{37}Cl	34.9689 36.9659	100 31.98
Bromine	79.909	^{79}Br ^{81}Br	78.9183 80.9163	100 97.3
Iodine	126.904	^{127}I	126.9045	100

These two forms of carbon give separate peaks in the mass spectrum, one mass unit apart. Thus the molecular ion peak is accompanied by an $[M+1]^+$ peak that corresponds to a molecular ion bearing one ^{13}C within its structure. If a mass spectrum of methane (CH_4) has a base peak at m/z 16 (100% relative intensity), it will also exhibit a peak at m/z 17 (1.1% intensity). If there are 10 carbons in a molecule, then the chances of the molecule containing one ^{13}C are 10 × 1.1, or 11%, and its M^+ peak will be accompanied by an $[M+1]^+$ peak that is 11% of the intensity of M^+. Thus, the relative abundance of the $[M+1]^+$ ions is related to the number of carbons in the structure times the natural abundance of ^{13}C. For a pure sample, the precision of this carbon count estimate can be within one or two carbons in a typical small organic compound. This is sufficient to distinguish whether one or more heavier atoms such as phosphorus or iodine may be accompanying the carbons within the structure.

For compounds containing S, Cl, and Br, there is a significant isotope peak present at $m/z = M^+ +2$ because the second-most abundant isotope of these elements contains two additional neutrons per atom (Table 8.1). Because these isotopes ^{34}S, ^{37}Cl, and ^{81}Br are much more abundant than ^{13}C, the $M^+ +2$ peak is much more likely to come from one of these elements than from a molecule containing two atoms of ^{13}C. Note that the likelihood of having two ^{13}C atoms within the same molecule is the natural abundance squared, times the number of carbons; for a 5-carbon compound this likelihood is $0.0111^2 × 5 = 0.0006$, or 0.06%, and that is negligible. Therefore the $M^+ +2$ peak is a useful diagnostic for the presence of S, Cl, or Br (**Figure 8.5**). On the other hand, the natural abundance of ^{81}Br is almost the same as its most abundant isotope, ^{79}Br. Therefore, for a compound having more than one Br, there would be significant peaks at both $M^+ +2$ and $M^+ +4$.

FIGURE 8.5

An electron impact (EI) mass spectrum of a compound containing an $M^+ +2$ isotope peak at m/z 92. We know that this compound contains Cl because the $M^+ +2$ ions have a relative abundance of 10% of the base peak, while M^+ ions at m/z 90 are 30% of the base peak. The ratio 10/30 = 0.33 corresponds closely to the 32% abundance of ^{37}Cl relative to ^{35}Cl.

Karty, J. *Organic Chemistry: Principles and Mechanisms*, 3rd ed.; W. W. Norton: New York, 2022; p 766.

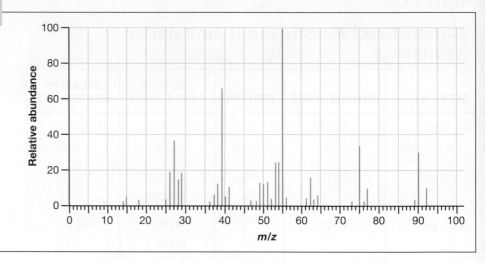

8.1E Determination of Molecular Formula

The first step to determining a molecular formula from an unknown compound is to find the molecular ion. For a pure sample of typical organic small molecules, the molecular ion is often found in a cluster of peaks at the high end of the observed m/z range. The largest peak in that cluster will generally be used to determine the molecular formula. There are some cautionary notes to mention, however. If higher energy ionization processes (EI or CI) are used, the molecular ion may appear small or even undetectable, depending on how readily the molecular ion fragments to smaller ions. If the sample is impure, higher molecular weight impurities can also confuse the matter. The ideal scenario is when the molecular ion is easily found as the base peak, which often occurs when electrospray ionization (ESI) is used, as its mild ionization conditions induce minimal fragmentation.

If you are using CI or ESI ionization, you will need to subtract the mass of H from $[M+H]^+$, or Na from $[M+Na]^+$, in order to calculate the mass of M^+ before proceeding to determine the molecular formula. If the molecular ion is visible in an EI mass spectrum it will be M^+ and no adjustment is needed. For typical organic compounds containing C, H, N, O, S, and halogen, here is a procedure you can use, starting with M^+:

1. Evaluate the $M^+ + 2$ peak for the presence of S, Cl, or Br.
   For S, the $M^+ + 2$ peak will be 4.4% of the M^+ peak.
   For Cl, the $M^+ + 2$ peak will be 32% of the M^+ peak.
   For Br, the $M^+ + 2$ peak will be 98% of the M^+ peak (nearly equal intensities).

2. Determine if the mass of M^+ is even or odd.
   If M^+ is even, there is an even number of nitrogens (0, 2, 4, etc.).
   If M^+ is odd, there is an odd number of nitrogens (1, 3, etc.).

3. Estimate the carbon count (C_n) from the intensity of the $M^+ + 1$ peak (relative to M^+).
   General formula: $n = $ (intensity of $M^+ + 1$/intensity of M^+) $\times$ 100/1.1

4. Add hydrogens and oxygens to the formula.
   Start with C_n plus any N, Cl, Br, or S.
   Add $2n + 2$ hydrogens.
   Add the minimum number of oxygens needed for the formula mass to exceed the observed m/z.
   Subtract hydrogens as needed to match the exact mass.

Solved Problem

Identify a possible formula for a pure compound that gives the following mass spectral data:

m/z	RELATIVE INTENSITY
86 (M^+)	10.00%
87	0.56%
88	0.04%

Solution

1. Because the $M^+ + 2$ peak is very small (only 0.4% relative to the M^+ peak), there is no Cl, Br, or S.

2. From M^+ we determine the molecular weight is 86 g/mol. An even or odd number for M^+ indicates the number of nitrogens is even or odd, respectively. In this case, m/z of M^+ is even (86), indicating an even number of nitrogens (0, 2, 4, ...).

3. The $M^+ + 1$ peak gives us an estimate of the carbon count:

$$\text{Number of carbons} = (0.56/10) \times 100/1.1 = 5$$

4. The formula C_5 by itself corresponds to 60 g/mol, so more atoms must be added. Start with C_5 and try various formulas to find one which adds up to 86, adding N, H, and O as necessary.
   Try $C_5N_2 = 60 + (2 \times 14) = 88$ g/mol (2 N gives MW > 86; must be 0 N)
   Try $C_5H_{12} = 60 + (1 \times 12) = 72$ g/mol (saturated, but MW too low; try adding O)
   Try $C_5H_{12}O = 88$ g/mol (MW is close; can adjust by adding unsaturation)
   Try $C_5H_{10}O = 86$ g/mol $\rightarrow$ SOLVED!

In this solved problem, $C_5H_{10}O$ is one possible formula for this compound. Because the number of carbons is not always very precise, you may wish to use the same procedure to determine formulas with carbon counts that are adjacent to C_5 (such as $C_4H_{10}N_2$, $C_4H_6O_2$, or C_6H_{14}). These can serve as alternative formulas that could potentially be distinguished by further information such as IR or NMR spectroscopy. As mass numbers get larger, there may be many more formulas that fit the mass spectrometry data.

8.1F Fragmentation and Structure

With ionization methods that impart high energies to the ions, especially EI and, to a lesser extent, CI, the ions that reach the detector include smaller fragments of the molecule in addition to the original molecular ion. Thus a mass spectrum generated in this way can be quite complex. Historically, this kind of spectrum has been analyzed carefully to determine formulas of various fragments, which can be used to reconstruct a possible structure. This is possible because the fragmentation can be predictable. It usually occurs at weaker bonds that can produce relatively stable cationic fragments. Therefore the masses of fragments can be correlated to the positions of various organic functional groups within the molecular structure, helping to illuminate its connectivity. A selection of common fragmentation processes is illustrated in **Figure 8.6**. However, for small molecules this kind of fragmentation analysis has been largely displaced by powerful NMR spectroscopy methods that give more direct information on structural connectivity.

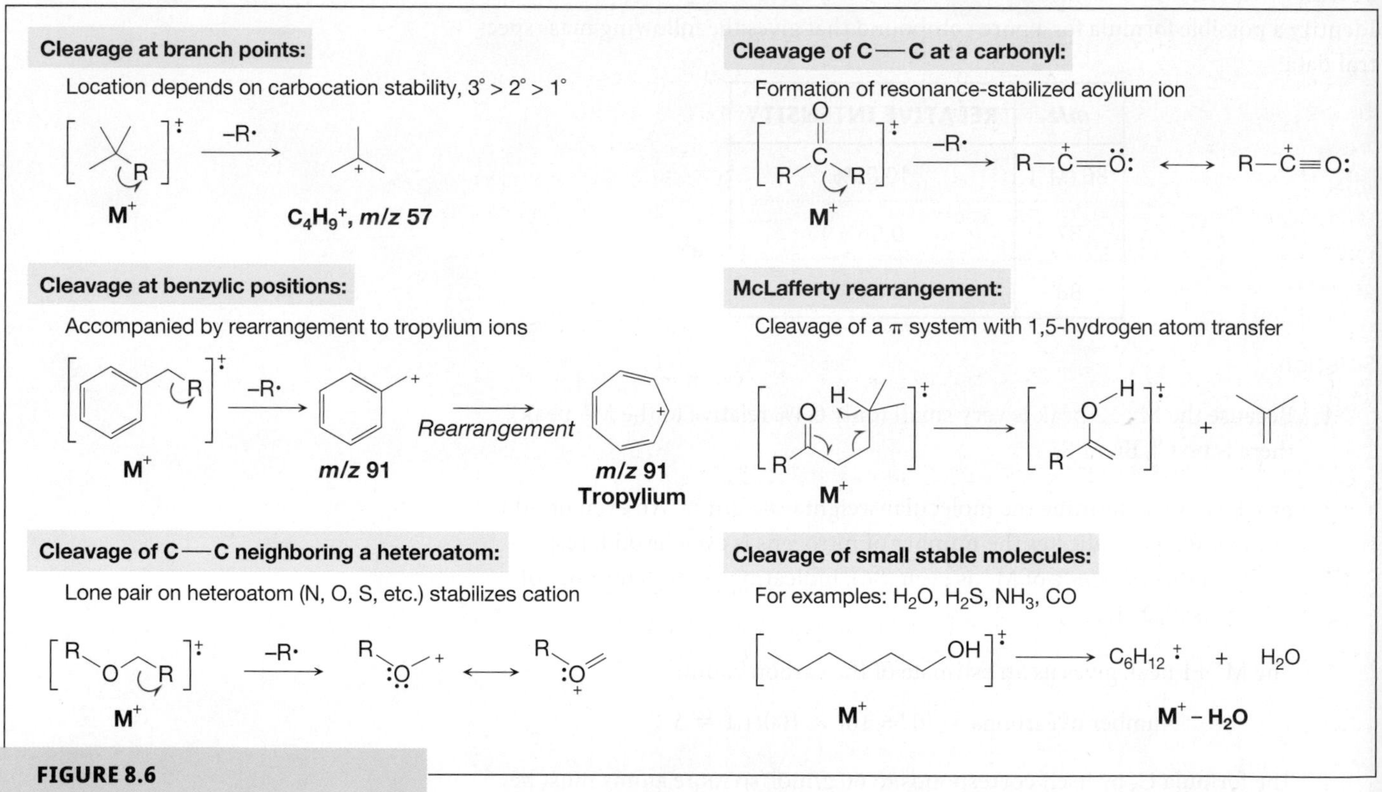

Cleavage at branch points:

Location depends on carbocation stability, 3° > 2° > 1°

$C_4H_9^+$, *m/z* 57

Cleavage at benzylic positions:

Accompanied by rearrangement to tropylium ions

m/z 91

Rearrangement

m/z 91
Tropylium

Cleavage of C—C neighboring a heteroatom:

Lone pair on heteroatom (N, O, S, etc.) stabilizes cation

Cleavage of C—C at a carbonyl:

Formation of resonance-stabilized acylium ion

McLafferty rearrangement:

Cleavage of a π system with 1,5-hydrogen atom transfer

Cleavage of small stable molecules:

For examples: H_2O, H_2S, NH_3, CO

C_6H_{12} + H_2O

$M^+ - H_2O$

FIGURE 8.6

Several common modes of fragmentation observed in electron-impact (EI) mass spectrometry.

8.1G High-Resolution Mass Spectrometry

Some mass spectrometers are able to differentiate compounds that would appear to have the same molecular weight, such as ethene (H_2C=CH_2) and dinitrogen (N_2). For both of these we would predict a M^+ of *m/z* 28 amu (atomic mass units) in the mass spectrometry methods discussed so far, which are considered low-resolution mass spectrometry. However, some instruments are capable of resolutions of 0.001 amu, 0.0001 amu, or even higher, and these allow for *high-resolution mass spectrometry* (HRMS). With this level of resolution, we must use very high-precision atomic masses of elements from the periodic table to calculate the m/z of the molecular ion. To do this, we use the "exact masses" of the most abundant isotopes (see Table 8.1) to calculate the "exact mass" m/z of the molecular ion composed of the most abundant isotopes. Using the exact masses 1.00783, 12.00000, and 14.0031 for hydrogen, carbon, and nitrogen, respectively, we can see that a high-resolution mass spectrum with resolution of 0.0001 could easily distinguish between ethene (exact mass 28.0313) and N_2 (exact mass 28.0062).

The high-resolution mass spectrum is a powerful way to determine the molecular formula of an unknown compound.

Worked Example

In an HRMS analysis of an unknown sample, a molecular ion M^+ of m/z 100.0890 was found. Is this compound 1,2-diaminocyclopentane (100 g/mol), cyclohexanol (100 g/mol), or heptane (100 g/mol)?

Solution

First, use the most abundant isotope exact masses from Table 8.1 to calculate the exact mass of the molecular ion in each case (**Table 8.2**). Then, compare the calculated M^+ exact masses to the observed experimental data. The smallest difference between the calculated and observed masses is with cyclohexanol (Δ 0.0001 amu), so we can say that this is likely the compound.

A final caveat about HRMS: In most cases this analysis technique gives information about identity, but with no indication of purity. A sample of 95% or 5% purity may yield a molecular ion peak of the same exact mass, and it may be difficult to know if the resulting data are from a minor impurity or from the major component in the sample. However, like other mass spectrometry methods, HRMS can be combined with gas chromatography (GC) or other separation techniques, so that HRMS data are obtained from each peak eluted during the chromatography. Information about identity and purity may then be available from a single analysis.

TABLE 8.2

Calculated Exact Masses (Worked Example)

COMPOUND, FORMULA	ISOTOPE EXACT MASSES	M⁺ EXACT MASS, CALCULATED
1,2-Diaminocyclopentane, $C_5H_{12}N_2$	^{12}C: 5 × 12.00000 = 60.00000 ^1H: 12 × 1.00783 = 12.09396 ^{14}N: 2 × 14.0031 = 28.0062	For $[C_5H_{12}N_2]^+$, m/z 100.1002
Cyclohexanol, $C_6H_{12}O$	^{12}C: 6 × 12.00000 = 72.00000 ^1H: 12 × 1.00783 = 12.09396 ^{16}O: 15.9949	For $[C_6H_{12}O]^+$, m/z 100.0889
Heptane, C_7H_{16}	^{12}C: 7 × 12.00000 = 84.00000 ^1H: 16 × 1.00783 = 16.12528	For $[C_7H_{16}]^+$, m/z 100.1253

| 8.2A | ## Elemental Ratios from Combustion By-Products |

Prior to any sort of spectroscopy, molecular formulas of organic compounds were determined by combustion and analysis of the combustion products, and this remains a valuable method today. Combustion of a precisely weighed sample of an organic compound at high temperatures with an oxygen source produces carbon dioxide (CO_2) and water (H_2O), and their amounts can be precisely measured. Similarly, if nitrogen gas (N_2) is produced by the combustion process, its amount can be measured. Together, these quantitative measurements of combustion gases allow the chemist to obtain a *CHN analysis* of a sample, which consists of a percentage by weight of the elements carbon, hydrogen, and nitrogen in the sample. Although there are variations on this method that can determine the percentages of sulfur, phosphorus, and other elements, CHN analysis is the preferred method for typical organic chemistry samples.

There are three main purposes of CHN analysis:

- Propose a complete or partial molecular formula of an unknown organic compound. Using the molecular weight, along with independent experimental evidence of other elements that may be present (such as halogen), CHN analysis can be translated into a molecular formula.
- Independently confirm a molecular formula suggested by other evidence. If a molecular formula has been proposed on the basis of other evidence, the calculated elemental ratios for that formula may be compared with the observed elemental ratios, and if they match, this confirms the formula.
- Confirm the purity of a known substance. For a known substance, if the calculated and observed elemental ratios are in good agreement, this confirms that the substance is of high purity, e.g., >95%.

| 8.2B | ## Historical Importance of Combustion Analysis |

Because combustion analysis is a very old technique that predates routine IR and NMR spectroscopy by at least a century, it is worth noting its importance in the development of organic and biological chemistry. In the early 1800s, the combustion elemental analysis method played a key role in rudimentary investigations of botany and physiology, leading to the classification of animal and plant substances into carbohydrates, lipids, and proteins based on their differing ratios of carbon, hydrogen, and nitrogen. By 1831, Justus von Liebig had refined combustion analysis to a level of precision sufficient to propose the beginnings of a chemical understanding of nutrition, respiration, and metabolism as presented in his 1842 book, *Animal Chemistry*.[1] Liebig popularized the general concept that physiological processes can be related to organic chemical reactions observed in the laboratory. Identifying the specific structures and reactions involved would take many decades of organic chemistry research efforts, but Liebig's proposals, born of combustion analysis, comprised a visionary leap that inspired such efforts.

---

[1] Liebig, J. *Animal Chemistry; or, Organic Chemistry in Its Applications to Physiology and Pathology*; Gregory, W., Ed.; Johnson Reprint Corp.: New York, 1964. (A facsim. of the Cambridge edition of 1842.)

In Liebig's version of the combustion analysis techniques, the combustion gases were trapped within tubes containing chemical adsorbents, and the change in mass of these tubes was measured to determine the amount of each gas.[2] This required combustion of significant amounts (0.5–1.0 g) of sample for accurate measurements. Advances made by Fritz Pregl in the early 1900s lowered the sample size to 2–4 mg. This was an important advance in the analysis of organic compounds, leading to a Nobel Prize awarded to Pregl in 1923. R. M. Wilstätter, another Nobel Laureate honored for his studies on natural pigment compounds, having isolated very small quantities of a pigment from 10,000 cattle ovaries, was able to use Pregl's improved combustion analysis technique to show that this compound was carotene, the same pigment found in carrots; thus demonstrating that pigments from the animal and plant kingdoms were chemically identical.[3]

With modern methods, about 0.5–1 mg sample sizes are introduced into analyzer instruments that automate the combustion process and exploit gas chromatography for precise measurement of the gaseous products. At least four significant figures are needed in the mass measurement of the sample in order to obtain meaningful data, so smaller samples require high-precision analytical balances (±0.1 mg or better). The instrument is calibrated using known compounds of high purity in order to ensure the output elemental ratios are accurate.

8.2C Proposing a Molecular Formula of an Unknown

A completely unknown compound cannot often be identified solely by elemental ratios, so additional information is generally required. Laboratory qualitative tests can determine the presence of certain elements such as sulfur or halogens. Sample history may suggest possible structures if the compound came from a known starting material and a routine type of organic reaction.

Worked Example

For an unknown compound, qualitative lab tests showed there was no halogen or sulfur present, and the following elemental mass ratios were determined by combustion analysis, in % by weight: C 58.53, H 4.09, N 11.38. Propose a molecular formula for this compound.

Solution

First, subtract all the C, H, N percentages from 100% in order to find the percentages of other elements present. We assume the remainder is oxygen since we are provided independent information that there is no halogen or sulfur present (**Table 8.3**). Divide the element percentages by their respective atomic masses to obtain the mole ratios of these elements. Next, convert these mole ratios to integers by dividing each of these by the smallest one to obtain relative amounts.

The results in the right-hand column all conform very closely to integers, so we can propose the molecular formula as $C_6H_5NO_2$. However, it should be noted the combustion analysis gives an empirical formula, and the same results would be obtained from, for example, $C_{12}H_{10}N_2O_4$.

TABLE 8.3

Formula from Mass % (Worked Example)

ELEMENT, MASS %	MOLE RATIO	INTEGER FORM
C 58.53%	58.53/12 = 4.88	4.88/0.813 = 6.00
H 4.09%	4.09/1 = 4.09	4.09/0.813 = 5.03
N 11.83%	11.38/14 = 0.813	0.813/0.813 = 1
O (100 − C, H, N) = 25.55%	25.55/16 = 1.60	1.60/0.813 = 1.97

[2]Holmes, F. L. Elementary Analysis and the Origins of Physiological Chemistry. *Isis* **1963**, *54* (1), 50–81.
[3]Pregl, F. Quantitative Micro Analysis of Organic Substances. Nobel Lecture, December 11, 1923. https://www.nobelprize.org/prizes/chemistry/1923/pregl/lecture (accessed May 2022).

If the problem-solving method above leads to numbers in the right-hand column that appear to include a half of an atom, this is nonsensical, as it is impossible to have a half of an atom. Multiply all elements by two to make them all integers in the formula. Similarly, if there are results such as 1.33 or 2.67, with 1/3 or 2/3 of an atom, then multiply all elements by three to make them integers.

8.2D | Confirmation of Molecular Formula

Often the chemist comes to a combustion analysis with a pure compound and some knowledge of its formula, and perhaps also the structure. Some of these situations may involve two or more alternative formulas/structures that seem consistent with whatever other information is at hand, such as IR or NMR spectroscopy. In such situations, the role of combustion analysis is to confirm one formula by comparison of predicted and observed elemental ratios. Typically, such comparisons are considered a match if the observed percentage of an element is within 0.4 of the calculated percentage.

8.2E | Assessment of Purity of a Known Compound

The presence of some impurities in a sample, such as solvents or leftover reactants, may be detected readily by NMR spectroscopy, and their amounts measured by integration. The presence of impurities (but not their identity) can also be detected if combustion analysis does not match the calculated ratios. If the difference between calculated and observed elemental percentage is more than 0.4, then the compound fails the assessment of purity.

In some cases impurities are undetectable by NMR. Combustion analysis takes on extra importance when such impurities could be present. For example, solvents without protons (e.g., CCl_4), various inorganic salts used in workups (e.g., NaCl), and adsorbents used in chromatography (e.g., silica gel) are all lacking signals in a ^{1}H NMR spectrum. These all happen to have low carbon percentages also, so they contribute to the mass of the sample, but do not produce CO_2 on combustion, making the carbon percentage appear lower than what is calculated for the pure compound. So a combustion analysis of a sample containing any of these impurities will not give a satisfactory CHN analysis; that is, the observed percentage of one or more elements will likely be more than 0.4% off the calculated percentage.

8.3 STRUCTURAL CLUES FROM MOLECULAR FORMULA

degrees of unsaturation >>
A measurement of the amount of hydrogen that is absent from a structure due to the presence of pi bonds or ring connections, in comparison with the saturated hydrocarbon formula (C_nH_{2n+2}). Each ring and/or pi bond introduced to a structure corresponds to a decrease of two hydrogens from the saturated hydrocarbon formula.

Whether the molecular formula is obtained from the mass spectrum, combustion analysis, or other means, there are clues within the formula that can be used to narrow down the range of possible structures. Some of these are related to the functional group. For example, if there is no nitrogen, you can rule out amines, amides, nitro compounds, and any other nitrogen-containing functional group. If there is only one oxygen, you can rule out carboxylic acids and esters. These simple considerations can get you on the right track to proposing a reasonable structure.

To refine the structural possibilities further, you can use a systematic assessment of the hydrogen content of the compound, which reflects whether it is saturated or unsaturated and to what degree (this is the organic chemistry version of oxidation states). The result of such an assessment is the **degrees of unsaturation** (also known as *double bond equivalents* or *index of hydrogen deficiency*) of the compound, and it gives information about how many π bonds or rings are present in the structure. To allow for one π bond or ring, a saturated hydrocarbon (C_nH_{2n+2}) requires the removal of two hydrogens from its structure in order for carbon to not exceed its valence of 4. A formula of C_nH_{2n} is missing two hydrogens in comparison with the saturated formula, and therefore has one degree of unsaturation.

For example, consider an unknown compound with the formula C_6H_{12}. The saturated hydrocarbon of the same carbon count has the formula C_nH_{2n+2} or in this case C_6H_{14}. Our unknown has two hydrogens fewer than the saturated compound, so we say it has one degree of unsaturation, and each possible structure we propose must have exactly one π bond or one ring in order to match the formula C_6H_{12}.

This type of analysis can be systematically applied to various formulas beyond hydrocarbons by subtracting the number of H's in the formula from the number of H's in the corresponding saturated hydrocarbon of the same carbon count, and then dividing the number of remaining H's by two. When heteroatoms (N, O, halogens, etc.) are present, some adjustments may be needed to account for the standard valences of the various heteroatoms:

- For compounds containing O, there is no adjustment to the H count.
- For every halogen X, add an H (then ignore the X).
- For every N, subtract an H (then ignore the N).

Worked Example

(a) Determine the degrees of unsaturation for the formula C_4H_4NOBr, and (b) propose a structure that is consistent with this formula.

1. The oxygen can be ignored.

2. For 1 N, subtract 1 H = C_4H_3.

3. For 1 X, add 1 H = C_4H_4.

4. Saturated hydrocarbon of the same carbon count = $C_nH_{2n+2} = C_4H_{10}$.

5. $H_{10} - H_4 = H_6$, and dividing by two we get 3 degrees of unsaturation.

With 3 degrees of unsaturation, any proposed structure must contain 3 π bonds, 3 rings, or any combination of rings and π bonds that totals 3. Note that triple bonds count as 2 π bonds. Two possible structures are shown in **Figure 8.7a**.

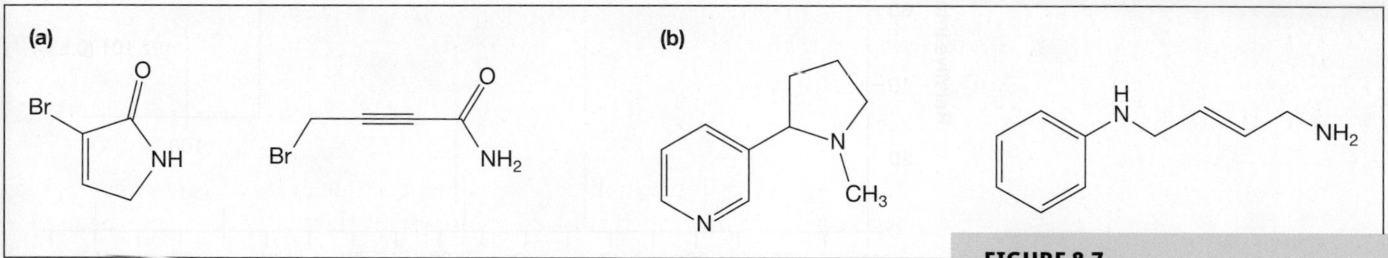

FIGURE 8.7

Possible structures corresponding to formulas (a) C_4H_4NOBr and (b) $C_{10}H_{14}N_2$.

Worked Example

Propose a structure having the formula $C_{10}H_{14}N_2$.

First, determine the degrees of unsaturation.

1. For 2 N, subtract 2 H = $C_{10}H_{12}$.

2. Saturated hydrocarbon = $C_nH_{2n+2} = C_{10}H_{22}$.

3. $H_{22} - H_{12} = H_{10}$, and dividing by two we get 5 degrees of unsaturation.

Second, consider the constraints on the proposed structure. With 5 degrees of unsaturation, any proposed structure must contain 5 π bonds, 5 rings, or any combination of rings and π bonds that totals 5. For the purposes of this analysis, benzene and other 6-membered aromatic rings are considered to have 3 π bonds, so a benzene ring would account for 4 degrees of unsaturation (one for the ring and the rest for the 3 π bonds). Two possible structures are shown in **Figure 8.7b**.

8.4 COMBINED SPECTROSCOPY PROBLEMS

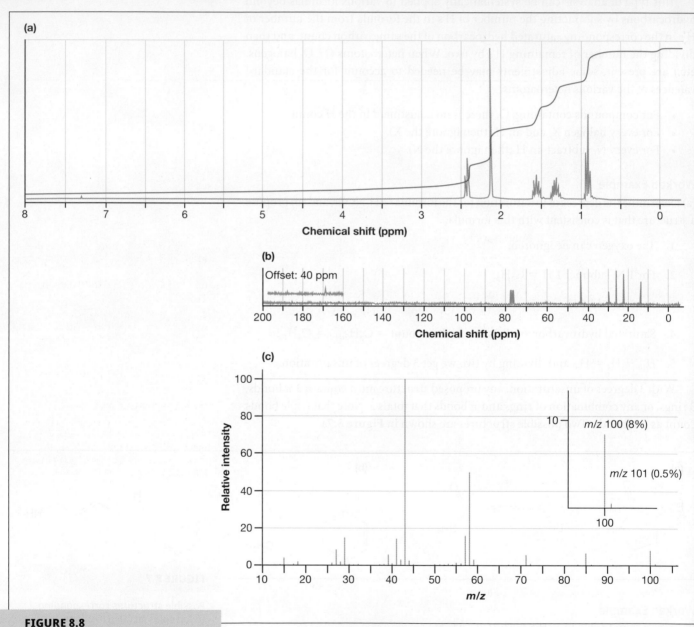

FIGURE 8.8

(a) ^{1}H NMR and (b) ^{13}C NMR data in CDCl$_3$ and (c) mass spectrum.

© Sigma-Aldrich Co. LLC. Reproduced with permission from Merck KGaA, Darmstadt, Germany and/or its affiliates. *Spectral Database for Organic Compounds, SDBSWeb*. National Institute of Advanced Industrial Science and Technology. https://sdbs.db.aist.go.jp/ (accessed August 2022). Reprinted by permission.

1. Identify the compound from the following ^{1}H NMR, ^{13}C NMR, mass spectrometry, and combustion analysis data. The ^{13}C and ^{1}H NMR spectra and mass spectrum are shown in **Figure 8.8**. Note that the chemical shift of the ^{13}C NMR peak shown in the offset at left is 40 ppm higher, at approximately 208 ppm.

   Combustion analysis, %: C, 71.91; H, 12.10

2. Identify the compound from the following ^{1}H NMR, ^{13}C NMR, IR, and HRMS data. The ^{13}C and ^{1}H NMR spectra are shown in **Figure 8.9**.
   IR spectrum, cm^{-1}: 2927, 2880, 1687
   High-resolution mass spectrum: m/z 99.0682 (M$^+$)

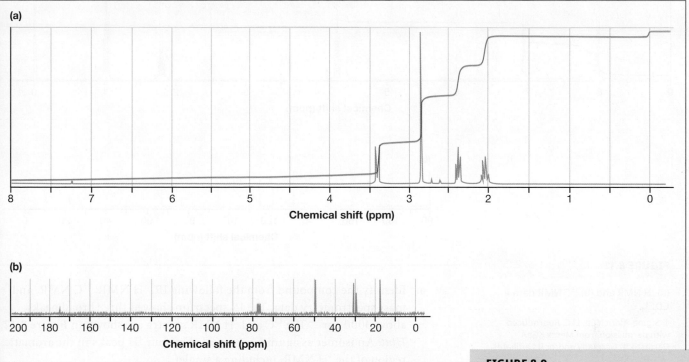

FIGURE 8.9

(a) ^{1}H NMR and (b) ^{13}C NMR data in CDCl$_3$.

© Sigma-Aldrich Co. LLC. Reproduced with permission from Merck KGaA, Darmstadt, Germany and/or its affiliates.

3. Identify the compound from the following ^{1}H NMR, ^{13}C NMR, IR, and combustion analysis data. The IR data are shown in **Figure 8.10**, and the ^{13}C and ^{1}H NMR spectra are in **Figure 8.11**.
   Combustion analysis, %: C, 66.64; H, 6.73

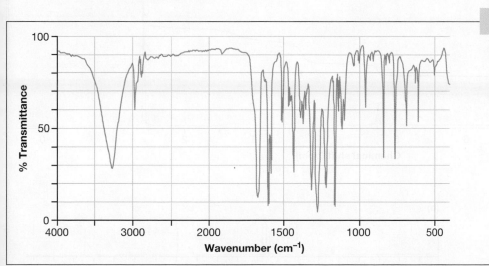

FIGURE 8.10

Infrared spectrum data.

Spectral Database for Organic Compounds, SDBSWeb. National Institute of Advanced Industrial Science and Technology. https://sdbs.db.aist.go.jp/ (accessed August 2022). Reprinted by permission.

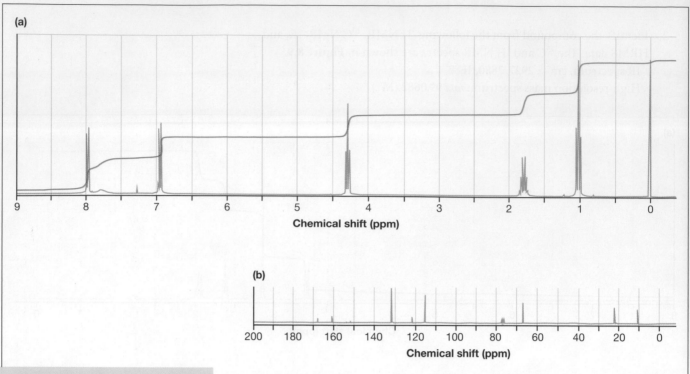

FIGURE 8.11

(a) ^{1}H NMR and (b) ^{13}C NMR data in CDCl$_3$.

© Sigma-Aldrich Co. LLC. Reproduced with permission from Merck KGaA, Darmstadt, Germany and/or its affiliates.

4. Identify the compound from the following IR, ^{1}H NMR, ^{13}C NMR, and mass spectrometry data. The IR spectrum showed three broad peaks above 3000 cm^{-1}. The ^{13}C and ^{1}H NMR spectra are shown in **Figure 8.12**. *Hint:* An isomer assignment is revealed by four ^{1}H peaks in the aromatic region of the ^{1}H NMR, including a singlet.

   Mass spectrum: *m/z* 137.08 (100%), 138.09 (8.7%)

   Combustion analysis, %: C, 70.02; H, 8.06; N, 10.20

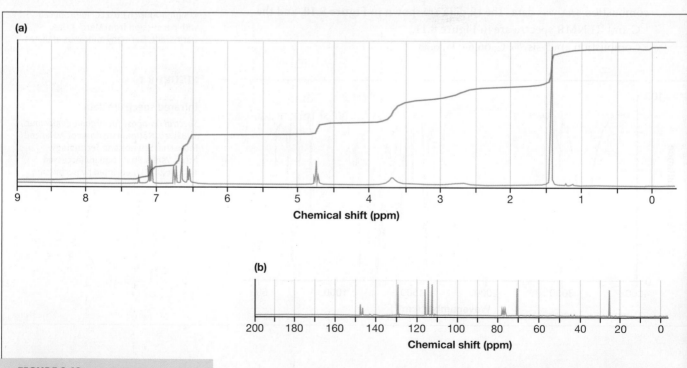

FIGURE 8.12

(a) ^{1}H NMR and (b) ^{13}C NMR data in CDCl$_3$.

Pouchert, C. J.; Behnke, J. *The Aldrich Library of ^{13}C and ^{1}H FT NMR Spectra*; Aldrich Chemical Company: Milwaukee, 1993.

Appendix 1

Infrared Spectroscopic Data

APPENDIX 1A

Typical Infrared Absorbances of Various Bonds

FREQUENCY (cm^{-1})	BOND	FUNCTIONAL GROUP
3500–3200 (s, b)	O—H	Alcohols, phenols
3400–3250 (m, sh)	N—H	Amines, amides
3300–2500 (m)	O—H	Carboxylic acids
3330–3270 (s, sh)	C≡C—H, C—H	Alkynes (terminal)
3100–3000 (m)	C—H, C (sp^2)	Aromatics, alkenes
3000–2850 (m)	C—H, C (sp^3)	Alkanes
2830–2695 (m)	H—C=O, C—H	Aldehydes
2260–2210 (v)	C≡N	Nitriles
2260–2100 (w)	C≡C—	Alkynes
1760–1665 (s)	C=O	Carboxylic acids, esters, amides, aldehydes, ketones
1680–1640 (m)	C=C	Alkenes
1600–1585 (m)	C—C (in-ring)	Aromatics
1320–1000 (s)	C—O	Alcohols, carboxylic acids, esters, ethers
850–515 (m)	C—Cl or C—Br	Alkyl halides

(s) = strong, (m) = medium, (w) = weak, (v) = variable intensity, (b) = broad, (sh) = sharp

Infrared Data for Representative Compounds with C＝O Bonds

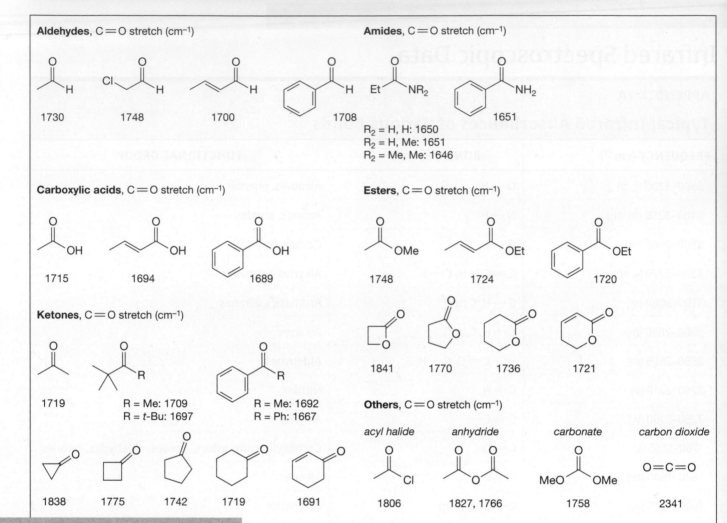

Aldehydes, C＝O stretch (cm⁻¹)

1730   1748   1700   1708

Amides, C＝O stretch (cm⁻¹)

1651

R₂ = H, H: 1650
R₂ = H, Me: 1651
R₂ = Me, Me: 1646

Carboxylic acids, C＝O stretch (cm⁻¹)

1715   1694   1689

Esters, C＝O stretch (cm⁻¹)

1748   1724   1720

1841   1770   1736   1721

Ketones, C＝O stretch (cm⁻¹)

1719   R = Me: 1709
        R = t-Bu: 1697

R = Me: 1692
R = Ph: 1667

Others, C＝O stretch (cm⁻¹)

acyl halide   *anhydride*   *carbonate*   *carbon dioxide*

1838   1775   1742   1719   1691

1806   1827, 1766   1758   2341

FIGURE A.1B

Silverstein, R. M.; Bassler, G. C.; Morrill, T. C. *Spectrometric Identification of Organic Compounds*, 4th ed.; Wiley: Hoboken, NJ, 1981. Crews, P.; Rodriguez, J.; Jaspars, M. *Organic Structure Analysis*; Oxford University Press: Oxford, 1998. Reich, H., University of Wisconsin, Chem 605 Handouts, personal communication, 2005. *AIST Spectral Database for Organic Compounds*. https://sdbs.db.aist.go.jp /sdbs/cgi-bin/cre_index.cgi (accessed March 2022).

Appendix 2

¹H NMR Spectroscopic Data

APPENDIX 2A

Typical Chemical Shift Ranges for Types of H

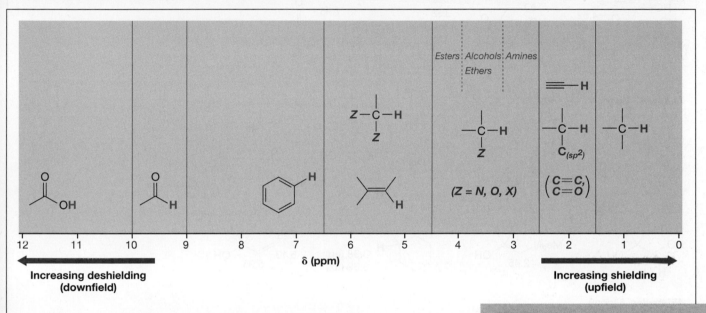

FIGURE A.2A

¹H NMR Chemical Shift Data for Representative Compounds

Acetals, δ (ppm)

Alcohols, δ (ppm)

Aldehydes, δ (ppm)

Alkenes, δ (ppm)

(continued)

¹H NMR Chemical Shift Data for Representative Compounds *(continued)*

Alkynes, δ (ppm)

Amides and Lactams, δ (ppm)

Amines, δ (ppm)

Aromatics, δ (ppm)

(continued)

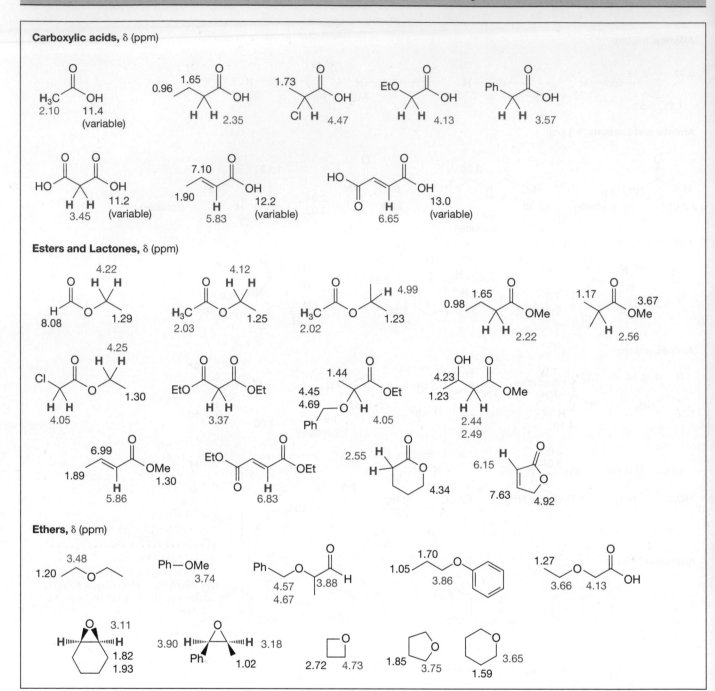

Carboxylic acids, δ (ppm)

Esters and Lactones, δ (ppm)

Ethers, δ (ppm)

(continued)

¹H NMR Chemical Shift Data for Representative Compounds *(continued)*

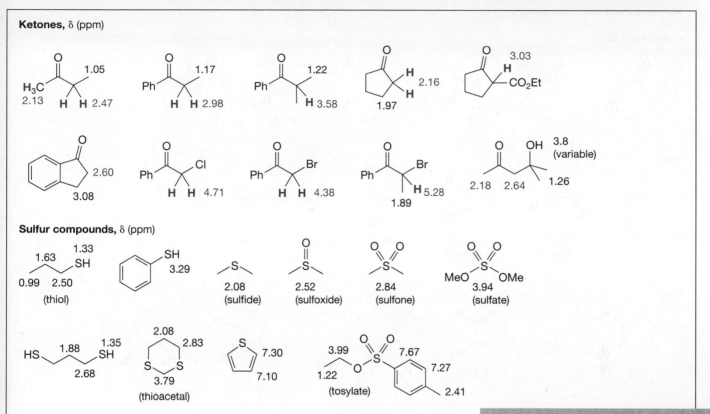

Ketones, δ (ppm)

Sulfur compounds, δ (ppm)

FIGURE A.2B

Data highlighted in blue are chemical shifts of hydrogens generally regarded as strongly diagnostic for the functional group. Numerical values should be regarded as approximations because solvent and concentration are not specified.

Silverstein, R. M.; Bassler, G. C.; Morrill, T. C. *Spectrometric Identification of Organic Compounds*, 4th ed.; Wiley: Hoboken, NJ, 1981. Crews, P.; Rodriguez, J.; Jaspars, M. *Organic Structure Analysis*; Oxford University Press: Oxford, 1998. Reich, H. University of Wisconsin, Chem 605 Handouts, personal communication, 2005. *AIST Spectral Database for Organic Compounds*. https://sdbs.db.aist.go.jp /sdbs/cgi-bin/cre_index.cgi (accessed March 2022).

Typical Ranges of Coupling Constants for Various ^1H–^1H Relationships

Vicinal, 3-Bond ^1H–^1H Coupling, J_{ab} (Hz)

- *alkenyl and dienyl*

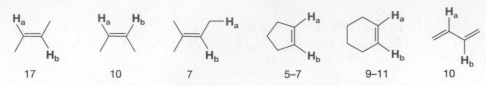

| 17 | 10 | 7 | 5–7 | 9–11 | 10 |

- *alkyl*

6–8 cis: 7 / trans: 6 cis: 6–9 / trans: 5–7 ax–ax: 12 ax–eq: 4 eq–eq: 4

- *aryl*

7.5 5.5 7.5 1.8 3.4

- *oxygen-containing*

5 2–3 6 2.5 4

Geminal, 2-Bond ^1H–^1H Coupling, J_{ab} (Hz)   **4-Bond ^1H–^1H Coupling, J_{ab} (Hz)**

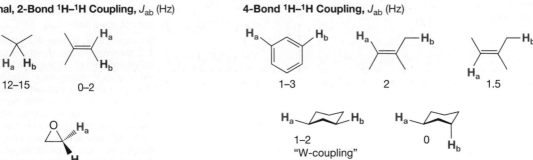

12–15   0–2   1–3   2   1.5

6   1–2 "W-coupling"   0

FIGURE A.2C

These are abbreviated and generalized structures; observation of signal splitting according to these coupling constants presumes that other substituents are present, rendering H_a and H_b nonequivalent.

Silverstein, R. M.; Bassler, G. C.; Morrill, T. C. *Spectrometric Identification of Organic Compounds*, 4th ed.; Wiley: Hoboken, NJ, 1981. Crews, P.; Rodriguez, J.; Jaspars, M. *Organic Structure Analysis*; Oxford University Press: Oxford, 1998.

Appendix 3

¹³C NMR Spectroscopic Data

APPENDIX 3A

Typical Chemical Shift Ranges for Types of C

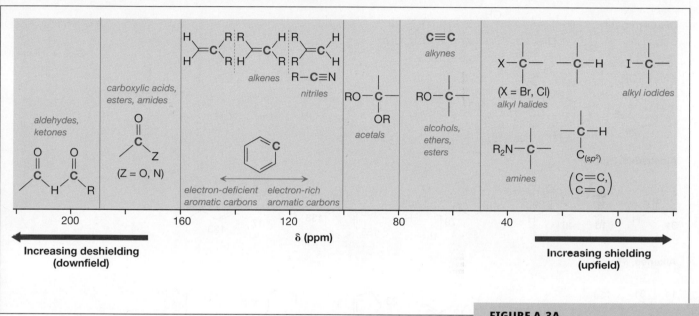

FIGURE A.3A

¹³C NMR Data for Representative Compounds

Acetals, δ (ppm)

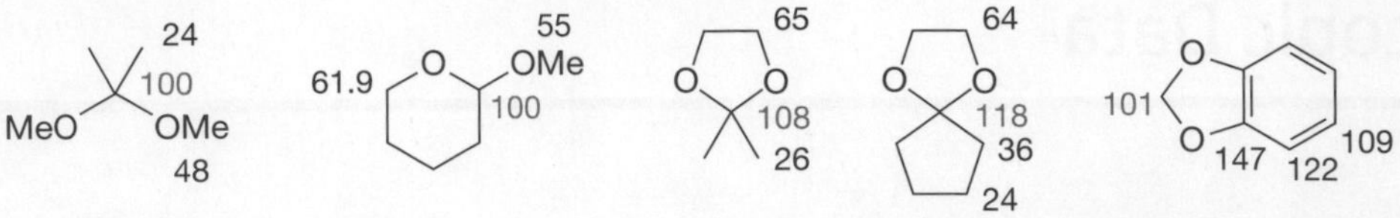

Alcohols, δ (ppm)

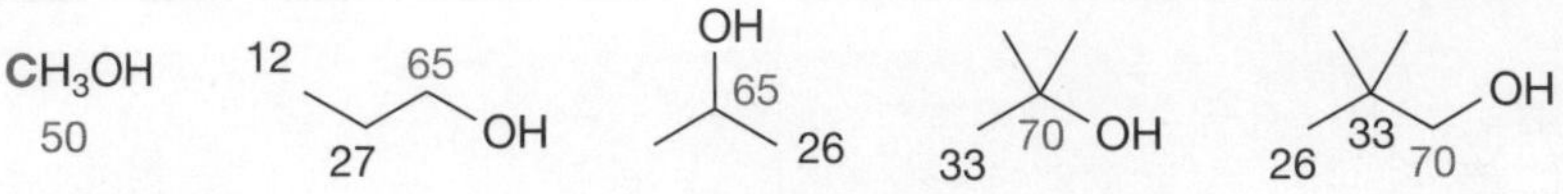

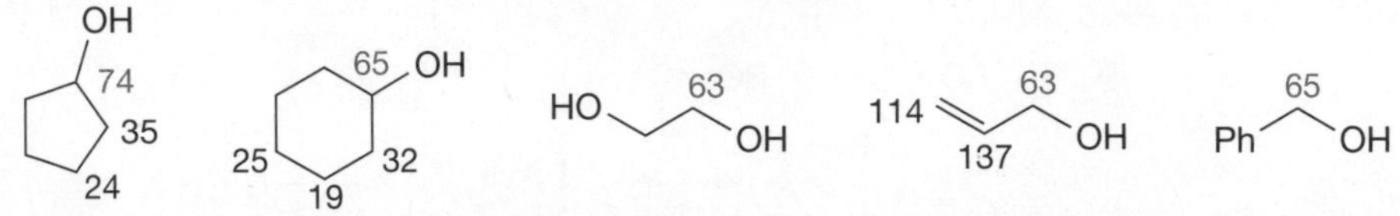

Aldehydes, δ (ppm)

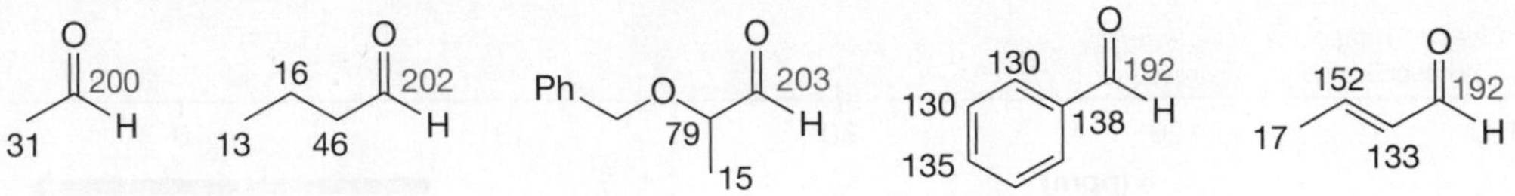

Alkenes, δ (ppm)

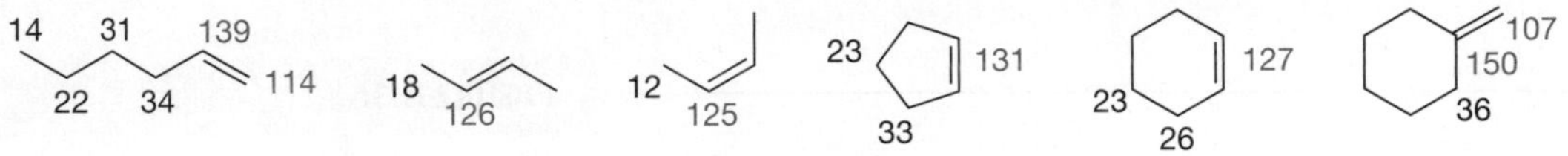

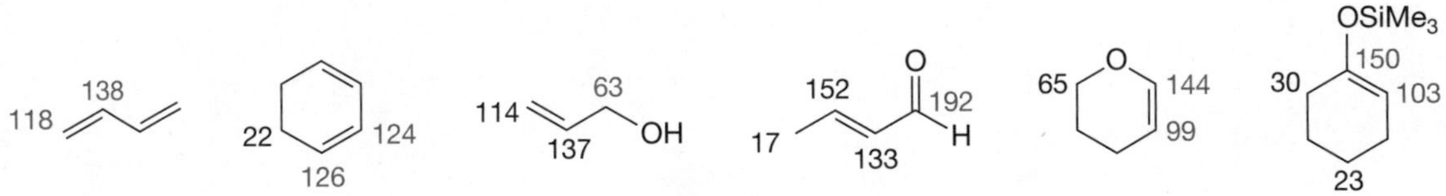

Alkynes, δ (ppm)

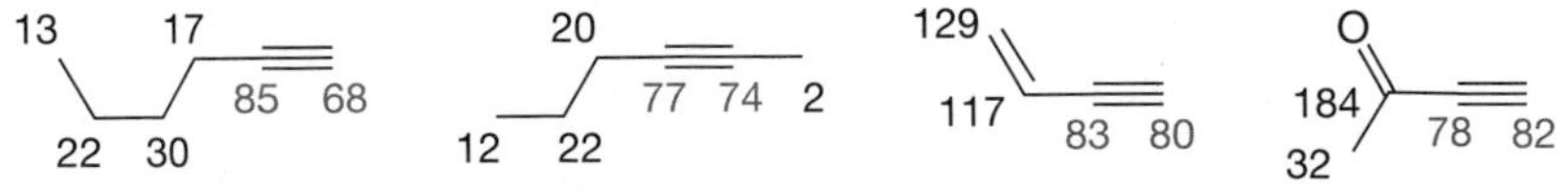

Amines, δ (ppm)

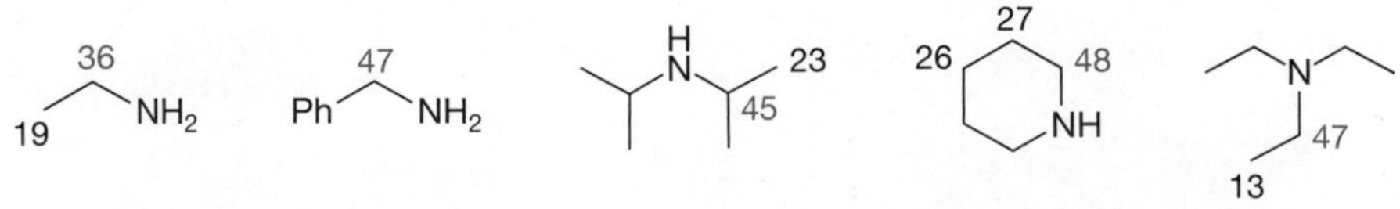

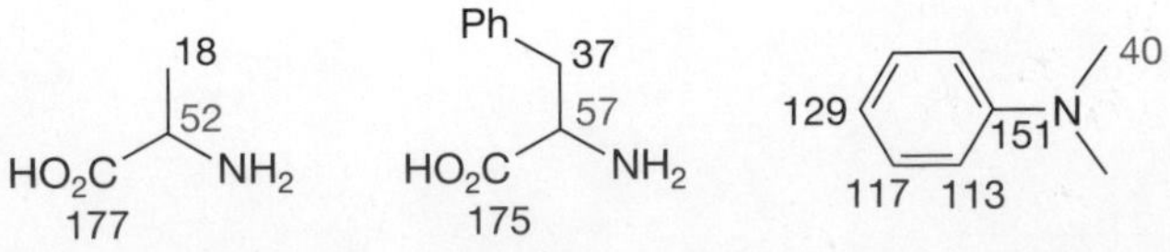

(continued)

^{13}C NMR Data for Representative Compounds *(continued)*

Amides and Lactams, δ (ppm)

Aromatics, δ (ppm)

Carboxylic acids, δ (ppm)

Esters and Lactones, δ (ppm)

(continued)

¹³C NMR Data for Representative Compounds *(continued)*

Ethers, δ (ppm)

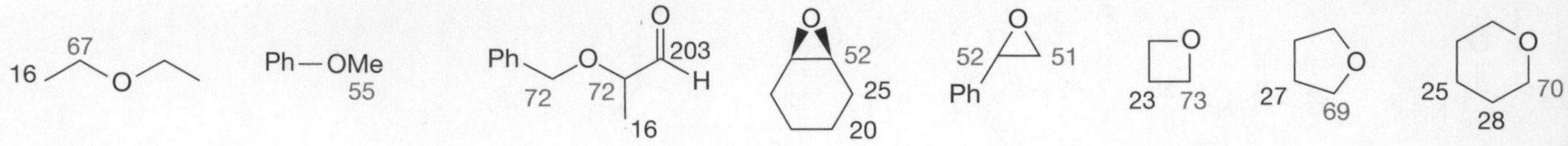

Ketones, δ (ppm)

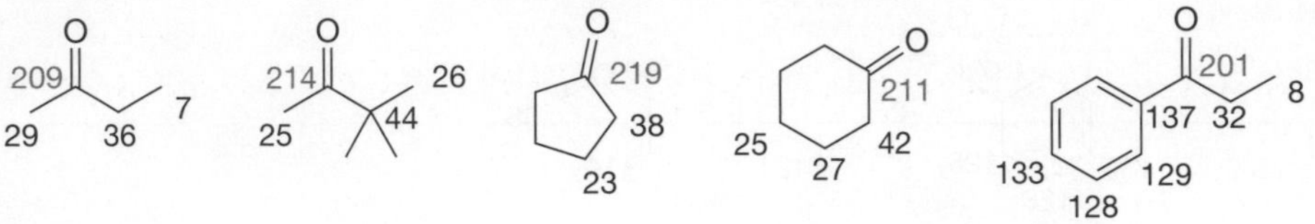

Sulfur compounds, δ (ppm)

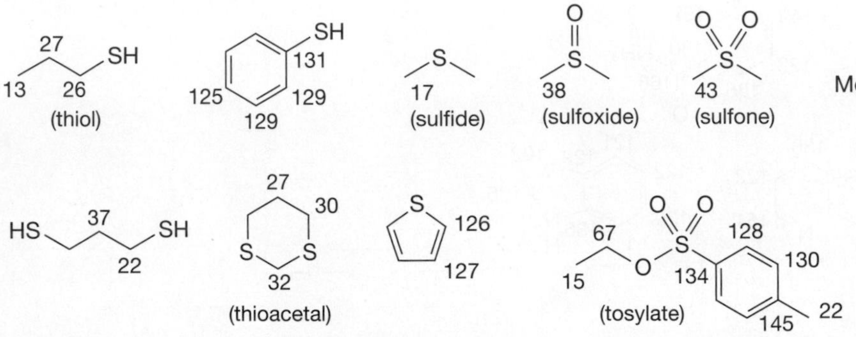

(thiol)   (sulfide)   (sulfoxide)   (sulfone)   (sulfate)

(thioacetal)   (tosylate)

FIGURE A.3B

Reich, H. University of Wisconsin, Chem 605 Handouts, personal communication, 2005. *AIST Spectral Database for Organic Compounds*. https://sdbs.db.aist.go.jp /sdbs/cgi-bin/cre_index.cgi (accessed March 2022).

Appendix 4

¹H NMR Data for Identification of Common Impurities

Observed ¹H NMR Chemical Shifts (δ, ppm) of Residual Solvent Impurity Peaks in Three Common NMR Solvents

	NMR SOLVENT		
IMPURITY	**CHLOROFORM-d, CDCl₃**	**ACETONE-d₆, (CD₃)₂C=O**	**DIMETHYLSULFOXIDE-d₆, (CD₃)₂S=O**
Water	1.56	2.84	3.33
Acetic acid	2.10	1.96	1.91
Acetone	2.17	2.09	2.09
tert-Butyl methyl ether	3.22, 1.19	3.13, 1.13	3.09, 1.11
Chloroform	7.26	8.02	8.32
Dichloromethane	5.30	5.63	5.76
Diethyl ether	3.48 (q), 1.21 (t)	3.41 (q), 1.11 (t)	3.38 (q), 1.09 (t)
Dimethylsulfoxide	2.62	2.52	2.54
Ethanol	3.72 (q), 1.32 (s, OH), 1.25 (t)	3.57 (q), 3.39 (s, OH), 1.12 (t)	4.63 (s, OH), 3.44 (q), 1.06 (t)
Ethyl acetate	4.12 (q), 2.05 (s), 1.26 (t)	4.05 (q), 1.97 (s), 1.20 (t)	4.03 (q), 1.99 (s), 1.17 (t)
n-Hexane	1.26 (m), 0.88 (t)	1.28 (m), 0.88 (t)	1.25 (m), 0.86 (t)
Methanol	3.49, 1.09 (s, OH)	3.31, 3.12 (s, OH)	4.01 (s, OH), 3.16
Pyridine	8.62 (m), 7.68 (m), 7.29 (m)	8.58 (m), 7.76 (m), 7.35 (m)	8.58 (m), 7.79 (m), 7.39 (m)
Tetrahydrofuran	3.76 (m), 1.85 (m)	3.63 (m), 1.79 (m)	3.60 (m), 1.76 (m)
Triethylamine	2.53 (q), 1.03 (t)	2.45 (q), 0.96 (t)	2.43 (q), 0.93 (t)

Unless noted, peak is observed as singlet (s). Other multiplicities are noted in parentheses: m = multiplet, q = quartet, t = triplet.
Gottlieb, H. E.; Kotlyar, V.; Nudelman, A. NMR Chemical Shifts of Common Laboratory Solvents as Trace Impurities. *J. Org. Chem.* **1997**, *62*, 7512–7515.

Appendix 5

Tables of Unknowns (Chapter 26): mp/bp and Derivative mp

APPENDIX 5A

Alcohols: Physical Properties and Solid Derivatives

COMPOUND	SYNONYM	mp (°C)	bp (°C)	3,5-DNB (°C)	PU (°C)
Methanol			65	108	47
Ethanol			78	93	52
2-Propanol	Isopropyl alcohol		82	123	88
2-Methyl-2-propanol	*tert*-Butyl alcohol	26	83	142	136
2-Propen-1-ol	Allyl alcohol		97	49	70
1-Propanol	*n*-Propyl alcohol		97	74	51
2-Butanol	*sec*-Butyl alcohol		98	76	65
2-Methyl-2-butanol	*tert*-Pentyl alcohol		102	116	42
2-Methyl-1-propanol	Isobutyl alcohol		108	87	86
3-Pentanol			115	101	48
1-Butanol	*n*-Butyl alcohol		118	64	63
2-Pentanol			119	62	
3-Methyl-3-pentanol			123	96	43
2-Methoxyethanol			124		113[a]
2-Methyl-3-pentanol			128	85	50
2-Chloroethanol			129	95	51
4-Methyl-2-pentanol			132	65	143
3-Hexanol			135	97	
1-Pentanol	*n*-Amyl alcohol		137	46	46
Cyclopentanol			140	115	132
2-Ethyl-1-butanol			146	51	
2,2,2-Trichloroethanol			151	142	87
1-Hexanol	*n*-Hexyl alcohol		156	58	42

(continued)

Alcohols: Physical Properties and Solid Derivatives *(continued)*

COMPOUND	SYNONYM	mp (°C)	bp (°C)	3,5-DNB (°C)	PU (°C)
Cyclohexanol			160	113	82
2-Hydroxymethylfuran	Furfuryl alcohol		170	80	45
1-Heptanol	*n*-Heptyl alcohol		176	47	65
2-Octanol			179	32	114
2-Ethyl-1-hexanol			185		61[a]
1-Octanol	*n*-Octyl alcohol		195	61	74
2-Nonanol			198	43	56[a]
3,7-Dimethyl-1,6-octadien-3-ol	Linalool		199		66
Benzyl alcohol			205	113	77
1-Nonanol	*n*-Nonyl alcohol		214	52	62
2-(4-Methyl-3-cyclohexenyl)isopropanol	α-Terpineol	36	221	78	112
1-Tetradecanol	Myristyl alcohol	39		67	74
2-Isopropyl-5-methylcyclohexanol	Menthol	41	212	158	111
1-Hexadecanol	Cetyl alcohol	49		66	73
2,2-Dimethyl-1-propanol	Neopentyl alcohol	56	113		144
4-Methylbenzyl alcohol		59	217	117	79
1-Octadecanol	Stearyl alcohol	59		77	79
Diphenylmethanol	Benzhydrol	68		141	139
4-Nitrobenzyl alcohol		93		157	
2-Hydroxy-2-phenylacetophenone	Benzoin	136			165
Cholesterol		148			168
2-Hydroxy-2,2-diphenylacetic acid	Benzilic acid	150		[b]	[b]

3,5-DNB = mp of 3,5-dinitrobenzoate derivative
PU = mp of phenylurethane derivative (except those with [a])
[a] Naphthylurethane derivative
[b] For derivatives, see carboxylic acids, Appendix 5F.

Aldehydes: Physical Properties and Solid Derivatives

COMPOUND	SYNONYM	mp (°C)	bp (°C)	DNP (°C)	SC (°C)
Propanal	Propionaldehyde		48	150	89
Propenal	Acrolein		52	165	171
2-Methylpropanal	Isobutyraldehyde		64	187	125
Butanal	Butyraldehyde		75	123	95
3-Methylbutanal	Isovaleraldehyde		92	123	107
Pentanal	Valeraldehyde		102	106	
2-Butenal	Crotonaldehyde		104	190	199
2-Ethylbutanal			117	134	99
Hexanal	Caproaldehyde		130	104	106
Heptanal	Heptaldehyde		153	108	109
Cyclohexanecarboxaldehyde			162		173
Furan-2-carboxaldehyde	Furfural		162	229	202
2-Ethylhexanal			163	114	254
Octanal			171	105	101
Benzaldehyde			179	237	222
Nonanal			185	100	100
Phenylethanal	Phenylacetaldehyde	33	195	121	153
2-Hydroxybenzaldehyde	Salicylaldehyde		197	248	231
4-Methylbenzaldehyde	Tolualdehyde		204	234	234
Decanal			207	104	102
3,7-Dimethyl-6-octenal	Citronellal		207	77	82
2-Chlorobenzaldehyde		11	213	213	225
3-Chlorobenzaldehyde		18	214	248	228
3-Bromobenzaldehyde			235		205
4-Methoxybenzaldehyde	*p*-Anisaldehyde		248	253	210
3,4-Methylenedioxybenzaldehyde	Piperonal	37		266	230
3,4-Dimethoxybenzaldehyde	Veratraldehyde	43		261	177
2-Nitrobenzaldehyde		44		265	256

(continued)

APPENDIX 5B

Aldehydes: Physical Properties and Solid Derivatives *(continued)*

COMPOUND	SYNONYM	mp (°C)	bp (°C)	DNP (°C)	SC (°C)
4-Chlorobenzaldehyde		48		254	230
4-Bromobenzaldehyde		57		257	228
3-Nitrobenzaldehyde		58		293	246
4-(*N*,*N*-Dimethylamino)benzaldehyde		74		325	222
4-Hydroxy-3-methoxybenzaldehyde	Vanillin	82		271	230
3-Hydroxybenzaldehyde		104		259	198
4-Nitrobenzaldehyde		106		320	221
4-Hydroxybenzaldehyde		116		280	224

DNP = mp of 2,4-dinitrophenylhydrazone derivative
SC = mp of semicarbazone derivative

Primary Amines: Physical Properties and Solid Derivatives

COMPOUND	SYNONYM	mp (°C)	bp (°C)	AA (°C)	BA (°C)	BSA (°C)	TSA (°C)	PT (°C)
tert-Butylamine			46		134			120
n-Propylamine			49		84	36	52	63
Allylamine			56			39	64	98
sec-Butylamine			63		76	70	55	101
Isobutylamine			69		57	53	78	82
n-Butylamine			77		42			65
Isopentylamine			95				65	102
1,2-Diaminoethane	Ethylenediamine		116	172	249	168	160	102
1,2-Diaminopropane			120	139	192		103	
n-Hexylamine			128		40	96		77
Cyclophexylamine			134	104	149	89		148
2-Aminomethylfuran	Furfurylamine		145	31	103			
n-Heptylamine			155					75
1,4-Diaminobutane		27	160					168
2-Hydroxyethylamine			171					135
Aniline			183	114	160	112	103	154
Benzylamine			184	60	105	88	116	156
1-Phenylethylamine	α-Phenethylamine		185	57	120			
2-Phenylethylamine	β-Phenethylamine		198	114	116	69		135
2-Methylaniline	*o*-Toluidine		199	112	143	124	108	136
3-Methylaniline	*m*-Toluidine		203	65	125	95	114	94
2-Chloroaniline			207	87	99	129	105	156
2-Ethylaniline			210	111	147			
4-Ethylaniline			216	94	151			104
2-Bromoaniline		31	229	99	116			146
1,6-Diaminohexane		42			155 (di)	154 (di)		
4-Methylaniline	*p*-Toluidine	45	200	153	158	120	117	141
2,5-Dichloroaniline		50		132	120			

(continued)

Primary Amines: Physical Properties and Solid Derivatives *(continued)*

COMPOUND	SYNONYM	mp (°C)	bp (°C)	AA (°C)	BA (°C)	BSA (°C)	TSA (°C)	PT (°C)
1-Naphthylamine		50		159	160	167	157	165
4-Methoxyaniline	*p*-Anisidine	58		127	154	95	114	154
3-Aminoaniline	*m*-Phenylenediamine	63		191	240	194	172	
2,4-Dichloroaniline		63		145	117	128	126	
4-Bromoaniline		66		167	204	134	101	148
4-Chloroaniline		70		179	192	121	95	152
2-Nitroaniline		71		92	94	104		142
4-Methyl-3-nitroaniline		77		148	172	160	164	145
2,4-Dibromoaniline		79		146	134		134	
2,6-Dibromoaniline		83		210				
Ethyl 4-aminobenzoate		89		110	148			
2-Methyl-3-nitroaniline		95		158				
4-Aminoacetophenone		106		167	205	128	203	
2-Methyl-5-nitroaniline		107		150		172		
4-Chloro-2-nitroaniline		118		104				
3-Aminophenol		122		101 (di)			157	156
4-Aminoaniline	*p*-Phenylenediamine	140		304	300	247	266	
4-Nitroaniline		147		210	199	139	191	
4-Aminoquinoline		154		178				
4-(Acetylamino)aniline	*p*-Aminoacetanilide	162		304				
2-Aminophenol		174		201	182	141	139	146
2,4-Dinitroaniline		180		120	202		219	
4-Aminophenol		184		150 (di)	234	125		150

AA = mp of acetamide derivative
BA = mp of benzamide derivative
BSA = mp of benzenesulfonamide derivative (Hinsberg product)
TSA = mp of *p*-toluenesulfonamide derivative
PT = mp of phenylthiourea derivative
di = derivatized at two sites

Secondary Amines: Physical Properties and Solid Derivatives

COMPOUND	SYNONYM	mp (°C)	bp (°C)	AA (°C)	BA (°C)	BSA (°C)	TSA (°C)	PT (°C)
Diethylamine			55		42	42	60	34
Diisopropylamine			86		67		81	
Pyrrolidine			88		48		133	
Piperidine			106		48	93	96	101
Di-*n*-propylamine			110			51		69
2-Methylpiperidine			116		45		55	
Morpholine			129		75	118	147	136
Diisobutylamine			139	86		55		113
Di-*n*-butylamine			160		49			86
N-Methylbenzylamine			185				95	
N-Methylaniline			192	102	63	79	94	87
N-Ethylaniline			205	54	60		87	89
Di-*n*-pentylamine			205					72
N-Methyl-*m*-toluidine			206	66				
N-Methyl-*o*-toluidine			207	55	66		120	
N-Methyl-*p*-toluidine			208	83	53	64	60	89
N-Benzylaniline		37		58	107	119		103
2-Nitro-*N*-methylaniline		37		70				
Indole		52			68			
Diphenylamine		54		101	180	124	141	152
3-Nitro-*N*-methylaniline		66		95	156	83		
Di-*p*-tolylamine		79		85	125			
4-Nitro-*N*-ethylaniline		96		118	98			
Piperazine		104	140	134	191		173	

AA = mp of acetamide derivative
BA = mp of benzamide derivative
BSA = mp of benzenesulfonamide derivative (Hinsberg product)
TSA = mp of *p*-toluenesulfonamide derivative
PT = mp of phenylthiourea derivative

Tertiary Amines: Physical Properties and Solid Derivatives

COMPOUND	SYNONYM	mp (°C)	bp (°C)	QAM (°C)
Triethylamine			89	280
Pyridine			115	117
2-Methylpyridine	2-Picoline		129	230
2,6-Dimethylpyridine	2,6-Lutidine		143	233
3-Methylpyridine	3-Picoline		144	92
Tri-*n*-propylamine			157	207
N,*N*-Dimethylbenzylamine			183	179
N,*N*-Dimethylaniline			193	228
Tri-*n*-butylamine			216	186
N,*N*-Diethylaniline			217	102
Quinoline			237	133
4-Bromo-*N*,*N*-dimethylaniline		55	264	185
N,*N*-Dibenzylaniline		70		135
4-Dimethylaminobenzaldehyde		74		161
Tribenzylamine		91		184

QAM = mp of quaternary ammonium methiodide derivative

Carboxylic Acids: Physical Properties and Solid Derivatives

COMPOUND	SYNONYM	mp (°C)	bp (°C)	AMIDE (°C)	ANILIDE (°C)	p-TOLUIDIDE (°C)
Formic acid	Methanoic acid	8	101	43	47	53
Acetic acid	Ethanoic acid	17	118	82	114	148
Propenoic acid	Acrylic acid	13	139	85	104	141
Propanoic acid	Propionic acid		141	81	103	124
2-Methylpropanoic acid	Isobutyric acid		154	128	105	104
Butanoic acid	Butyric acid		162	115	95	72
2-Methylpropenoic acid	Methacrylic acid	16	163	102	87	
2,2-Dimethylpropanoic acid	Pivalic acid	35	164	178	127	
3-Methylbutanoic acid	Isovaleric acid		176	135	109	109
Pentanoic acid	Valeric acid		186	106	63	70
2-Methylpentanoic acid			186	79	95	80
2-Chloropropanoic acid			186	80	92	124
Dichloroacetic acid			194	98	118	153
Hexanoic Acid	Caproic acid		205	101	95	75
2-Bromopropanoic acid		24	205	123	99	125
Octanoic acid	Caprylic acid	16	237	107	57	70
4-Oxopentanoic acid	Levulinic acid	33	246	108	102	108
Nonanoic acid		12	254	99	57	84
Decanoic acid		32	268	108	70	78
Lauric acid	Dodecanoic acid	43		98	76	
3-Phenylpropanoic acid	Hydrocinnamic acid	48		105	98	135
Bromoacetic acid		50		91	131	
Tetradecanoic acid	Myristic acid	54		103	84	93
Trichloroacetic acid		57	198	141	97	113
Hexadecanoic acid	Palmitic acid	62		106	90	98
Chloroacetic acid		63	189	121	137	162
Octadecanoic acid	Stearic acid	69		109	95	102

(continued)

Carboxylic Acids: Physical Properties and Solid Derivatives *(continued)*

COMPOUND	SYNONYM	mp (°C)	bp (°C)	AMIDE (°C)	ANILIDE (°C)	*p*-TOLUIDIDE (°C)
trans-2-Butenoic acid	Crotonic acid	72		158	118	132
Phenylacetic acid		77		156	118	136
4-Methoxyphenylacetic acid		87		189		
3,4-Dimethoxyphenylacetic acid		97		147		
Pentanedioic acid	Glutaric acid	98		176 (di)	224 (di)	218 (di)
2-Methoxybenzoic acid	*o*-Anisic acid	100		129	131	
2-Methylbenzoic acid	*o*-Toluic acid	104		142	125	144
3-Methoxybenzoic acid	*m*-Anisic acid	107		136		
3-Methylbenzoic acid	*m*-Toluic acid	111		94	126	118
4-Bromophenylacetic acid		117		194		
Benzoic acid		122		130	163	158
2-Benzoylbenzoic acid		127		165	195	
cis-Butenedioic acid	Maleic acid	130		260 (di)	187 (di)	142 (di)
2-Furoic acid		133		143	124	170
(*E*)-3-Phenylprop-2-enoic acid	*trans*-Cinnamic acid	133		147	153	168
2-Acetylsalicylic acid	Aspirin	138		138	136	
2-Chlorobenzoic acid		140		139	118	131
3-Nitrobenzoic acid		140		143	155	162
4-Chloro-2-nitrobenzoic acid		142		172		
2-Nitrobenzoic acid		146		176	155	
Diphenylacetic acid		148		167	180	172
2-Hydroxy-2,2-diphenylacetic acid	Benzilic acid	150		154	175	190
Adipic acid	Hexanedioic acid	152		220	235	
3-Chlorobenzoic acid		156		134	123	
2,4-Dichlorobenzoic acid		158		194		
2-Hydroxybenzoic acid	Salicylic acid	158		142	136	156
3,4-Dimethylbenzoic acid		165		130	104	

(continued)

Carboxylic Acids: Physical Properties and Solid Derivatives *(continued)*

COMPOUND	SYNONYM	mp (°C)	bp (°C)	AMIDE (°C)	ANILIDE (°C)	*p*-TOLUIDIDE (°C)
2-Chloro-5-nitrobenzoic acid		166		178		
Tartaric acid		169		196 (di)	264 (di)	
4-Methylbenzoic acid	*p*-Toluic acid	180		160	145	160
4-Methoxybenzoic acid	*p*-Anisic acid	184		167	169	186
Butanedioic acid	Succinic acid	188		260 (di)	230 (di)	255 (di)
4-Ethoxybenzoic acid		198		202	170	
3-Hydroxybenzoic acid		201		170	157	163
3,5-Dinitrobenzoic acid		202		183	234	
3,4-Dichlorobenzoic acid		209		133		
Benzene-1,2-dicarboxylic acid	Phthalic acid	210		220 (di)	253 (di)	201 (di)
4-Hydroxybenzoic acid		214		162	197	204
4-Nitrobenzoic acid		240		201	211	204
4-Chlorobenzoic acid		242		179	194	
4-Bromobenzoic acid		251		190	197	

Amide = mp of primary amide derivative (via acyl chloride with ammonia)
Anilide = mp of *N*-phenylamide derivative (via acyl chloride with aniline)
p-Toluidide = mp of *N*-(*p*-tolyl)amide derivative (via acyl chloride with *p*-toluidine)
di = derivatized at two sites

Esters: Physical Properties and Solid Derivatives

COMPOUND	SYNONYM	mp (°C)	bp (°C)	NBA (°C)	3,5-DNB (°C)	ACID (°C)
Methyl formate	Methyl methanoate		32	60	108	
Ethyl formate	Ethyl methanoate		54	60	93	
Methyl acetate	Methyl ethanoate		57	61	108	
Ethyl acetate	Ethyl ethanoate		77	61	93	
Methyl propanoate	Methyl propionate		80	47	108	
Isopropyl acetate	Propan-2-yl ethanoate		88	61	123	
Methyl 2-methylpropanoate	Methyl isobutyrate		93	88	108	
Ethyl propanoate	Ethyl propionate		99	47	93	
Propyl acetate	Propyl ethanoate		102	61	74	
Methyl butanoate	Methyl butyrate		102	38	108	
Ethyl 2-methylpropanoate	Ethyl isobutyrate		110	88	93	
2-Butyl acetate	sec-Butyl acetate		112	61	76	
2-Methyl-1-propyl acetate	Isobutyl acetate		118	61	87	
Ethyl butanoate	Ethyl butyrate		122	38	93	
Propyl propanoate	n-Propyl propionate		123	47	74	
Butyl acetate			126	61	64	
Methyl chloroacetate			132		108	63
3-Methylbutan-1-yl acetate	Isoamyl acetate		142	61	61	
Ethyl chloroacetate			145		93	63
Pentyl acetate	n-Amyl acetate		149	61	46	
Ethyl 2-hydroxypropanoate	Ethyl lactate		154	48	93	
Ethyl hexanoate	Ethyl caproate		168	55	93	
Hexyl acetate			169	61	58	
Methyl 3-oxobutanoate	Methyl acetoacetate		170	100	108	
Dimethyl malonate			180	142 (di)	108	
Ethyl 3-oxobutanoate	Ethyl acetoacetate		181	100	93	
Diethyl oxalate			185	221 (di)	93	
Heptyl acetate			192	61	47	

(continued)

Esters: Physical Properties and Solid Derivatives *(continued)*

COMPOUND	SYNONYM	mp (°C)	bp (°C)	NBA (°C)	3,5-DNB (°C)	ACID (°C)
Phenyl acetate			197	61	146	
Methyl benzoate			199	106	108	122
Ethyl octanoate	Ethyl caprylate		207	66	93	
2-Methylphenyl acetate	*o*-Tolyl acetate		208	61	135	
3-Methylphenyl acetate	*m*-Tolyl acetate		212	61	165	
Ethyl benzoate			212	106	93	122
4-Methylphenyl acetate	*p*-Tolyl acetate		213	61	189	
Methyl 2-methylbenzoate	Methyl *o*-toluate		215		108	104
Methyl 3-methylbenzoate	Methyl *m*-toluate		215	76	108	111
Benzyl acetate			217	61	113	
Methyl phenylacetate			220	122	108	77
Methyl 4-methylbenzoate	Methyl *p*-toluate	33	223	133	108	180
Methyl 2-hydroxybenzoate	Methyl salicylate		224		108	158
Ethyl phenylacetate			228	122	93	77
Methyl (2*E*)-3-phenylprop-2-enoate	Methyl cinnamate	36		226	108	133
Phenyl 2-hydroxybenzoate	Phenyl salicylate	42			146	158
Methyl 4-chlorobenzoate		44		163	108	242
Ethyl 4-nitrobenzoate		56		142	93	240
Phenyl benzoate		69		106	146	122
Methyl 3-nitrobenzoate		78		101	108	140
Methyl 4-bromobenzoate		81		168	108	251
Methyl 4-nitrobenzoate		94		142	108	240

NBA = mp of *N*-benzylamide derivative of the acyl component
3,5-DNB = mp of 3,5-dinitrobenzoate derivative of the alcohol component
Acid = mp of carboxylic acid obtained via saponification

Ketones: Physical Properties and Solid Derivatives

COMPOUND	SYNONYM	mp (°C)	bp (°C)	DNP (°C)	SC (°C)
2-Propanone	Acetone		56	126	187
2-Butanone	Methyl ethyl ketone		80	117	146
3-Methyl-2-butanone	Isopropyl methyl ketone		94	120	112
2-Pentanone			101	143	112
3-Pentanone			102	156	138
3,3-Dimethyl-2-butanone	Pinacolone, *tert*-butyl methyl ketone		106	125	157
4-Methyl-2-pentanone			117	95	132
2,4-Dimethyl-3-pentanone	Diisopropyl ketone		124	95	160
3-Hexanone			125	130	113
2-Hexanone			128	106	121
Cyclopentanone			130	146	210
2,4-Pentanedione	Acetylacetone		139	209	209 (di)
4-Heptanone			144	75	132
3-Heptanone			148		101
2-Heptanone			151	89	123
Cyclohexanone			156	162	166
3-Octanone			167		117
2-Octanone			173	58	122
Cycloheptanone			181	148	163
Ethyl 3-oxobutanoate	Ethyl acetoacetate		181	93	129
5-Nonanone			186		90
3-Nonanone			187		112
2-Nonanone			195		118
Acetophenone	Methyl phenyl ketone	20	202	238	198
1-Phenyl-2-propanone	Phenylacetone	27	216	156	198
1-Phenyl-1-propanone	Propiophenone	21	218	191	173
Carvone			225	193	142
1-Phenyl-2-butanone			226		135

(continued)

Ketones: Physical Properties and Solid Derivatives *(continued)*

COMPOUND	SYNONYM	mp (°C)	bp (°C)	DNP (°C)	SC (°C)
4-Methylacetophenone	Methyl *p*-tolyl ketone	28	226	258	205
2-Undecanone			231	63	122
4-Chlorophenyl-1-propanone	4-Chloropropiophenone	36			176
4-Methoxyphenyl-1-ethanone	4-Methoxyacetophenone	38		220	198
1-Indanone		41		258	233
Benzophenone	Diphenyl ketone	48		238	164
4-Bromophenyl-1-ethanone	4-Bromoacetophenone	51		230	208
1,2-Diphenylethanone	Benzyl phenyl ketone	60		204	148
1,1-Diphenyl-2-propanone	1,1-Diphenylacetone	61			170
4-Chlorobenzophenone		76			185
4-Bromobenzophenone		82		230	
Fluorenone		83		234	283
4-Hydroxyphenyl-1-ethanone	4-Hydroxyacetophenone	109		210	199
2-Hydroxy-2-phenylacetophenone	Benzoin	136		245	206
Camphor, (+)		179	205	177	237

DNP = mp of 2,4-dinitrophenylhydrazone derivative
SC = mp of semicarbazone derivative
di = derivatized at two sites

Phenols: Physical Properties and Solid Derivatives

COMPOUND	SYNONYM	mp (°C)	bp (°C)	NU (°C)	BROMINATION (°C)	OTHER (°C)
2-Chlorophenol			176	120	Mono 48, di 76	
2-Methylphenol	o-Cresol	32	191	142	Di 56	
3-Methylphenol	m-Cresol		203	128	Tri 84	
2-Methoxyphenol	Guaiacol	32	204	118	Tri 116	
2,4-Dimethylphenol		23	212	135		
Phenyl 2-hydroxybenzoate	Phenyl salicylate	42			Mono 113, di 128	Phenylurethane 112
4-Chlorophenol		43	217	166	Mono 33, di 90	
2,4-Dichlorophenol		45	210		Mono 68	
4-Ethylphenol		45	219	128		
2-Isopropyl-5-methylphenol	Thymol	51	234	160	Mono 55	
4-Bromophenol		64	238	169	Tri 95	
3,5-Dimethylphenol		68	220	109	Tri 166	
2,5-Dimethylphenol		75	212	173	Tri 178	
4-Hydroxy-3-methoxybenzaldehyde	Vanillin	82				(see Aldehydes)
1-Naphthol		96		152	Di 105	
2-Hydroxyphenol	Catechol	105		175	Tetra 192	
3-Hydroxyphenol	Resorcinol	109			Tri 112	
4-Hydroxyphenyl-1-ethanone	4-Hydroxyacetophenone	109				Phenylurethane 154
4-Nitrophenol		112		150	Di 142	
4-Hydroxybenzaldehyde		116			Mono 125	Phenylurethane 136
2-Naphthol		121		157	Mono 84	
1,2,3-Trihydroxybenzene	Pyrogallol	133			Di 158	
2-Hydroxybenzoic acid	Salicylic acid	158				Amide 142
3-Hydroxybenzoic acid		201				Amide 170
4-Hydroxybenzoic acid		214				Amide 162

NU = mp of naphthylurethane derivative
Bromination = mp of arene bromination products, with number of Br substitutions

Appendix 6

Bond Strengths of Representative Organic Compounds

C—H BONDS, HYDROCARBONS		C—H BONDS, OTHER FUNCTIONAL GROUPS		C—C BONDS		C—X BONDS	
COMPOUND, BOND	STRENGTH (kJ/mol)	COMPOUND, BOND	STRENGTH (kJ/mol)	COMPOUND, BOND	STRENGTH (kJ/mol)	COMPOUND, BOND	STRENGTH (kJ/mol)
H_3C—H	439	⌒OH / H	401	H_3C—CH_3	377	H_3C—Cl	350
⌒⌒⌒H	421			CH_3	369	Cl_3C—Cl	289
⌒ H	411	OH / H	384	CH_3	364	Cl	355
⋋ H	400	OH / H	341	⬡—CH_3	434	Cl	352
≡—H	558	⌒O⌒	389	⬡—CH_3	325	⬡—Cl	406
≡⌒H	384	O (furan) H	385	⬡—⬡	493	⬡—Cl	300
⌒H	464	O propanal H	375	⌒⌒OH	357	O / Cl (acyl)	353
⌒⌒H	369	O H	384	H_3C—H (acetaldehyde)	355	Cl—OH (acid)	311
⌒⋋H	351	O H	404	H_3C—CH_3 (acetone)	352	H_3C—Br	294
⌒⋋H	333	O H	386	Ph—CH_3	414	H_3C—I	232
▷—H	445	O H OH	399	O OH	385	H_3C—OH	385
◇—H	409	O H OEt	402	H_2N—OH	349	H_3C—OCH_3	352
⬠—H	400	H_2N—H	393			H_3C—NH_2	356
⬡—H	416	H_2N—H	377	**O—H BONDS**		**O—O, N—N, S—S BONDS**	
⬡—H	466	EtHN—H	371	HO—H	497	HO—OH	210
⬡—⌒H	376	Et_2N—H	380	EtO—H	435	EtO—OEt	166
⬡—⋋H	348			PhO—H	363	t-BuO—OH	186
				HOO—H	366	H_2N—NH_2	277
				ONO—H	330	PhS—SPh	214

Strength = Bond dissociation enthalpy
To convert data to kcal/mol, divide by 4.184 kJ/kcal.

Appendix 7

Typical pK_a Values of Representative Organic Compounds

Values of pK_a for Various Acids[a]

ACID	CONJUGATE BASE	pK_a	ACID	CONJUGATE BASE	pK_a
Trifluoromethanesulfonic acid (TfOH)		−13	Trichloroethanoic acid (Trichloroacetic acid)		0.77
HI Hydroiodic acid	I⁻	−10	Chloroethanoic acid (Chloroacetic acid)		2.87
Sulfuric acid		−9	HF Hydrofluoric acid	F⁻	3.2
HBr Hydrobromic acid	Br⁻	−9	Methanoic acid (Formic acid)		3.75
HCl Hydrochloric acid	Cl⁻	−7	Benzoic acid		4.2
p-Toluenesulfonic acid (TsOH)		−2.8	Ethanoic acid (Acetic acid)		4.75
Methanesulfonic acid (MsOH)		−2	Pyridinium ion		5.2
H_3O^+ Hydronium ion	H_2O	0.0[b]	Carbonic acid		6.3
Trifluoroethanoic acid (Trifluoroacetic acid)		0.0			

(continued)

Values of pK_a for Various Acids[a] (continued)

ACID	CONJUGATE BASE	pK_a	ACID	CONJUGATE BASE	pK_a
Thiophenol		6.6	2,2,2-Trifluoroethanol		12.4
4-Nitrophenol		7.2	Diethyl propanedioate (Diethyl malonate)		13.5
H_2S Hydrogen sulfide	HS^-	7.2	H_2O Water	HO^-	14.0[b]
2,4-Pentanedione		8.9	2-Chloroethanol		14.3
$N{\equiv}C{-}H$ Hydrocyanic acid	$N{\equiv}C^-$	9.2	Pyrrole		15
H_4N^+ Ammonium ion	NH_3	9.4	CH_3OH Methanol	CH_3O^-	15.5
$(CH_3)_3\overset{+}{N}H$ Trimethylammonium ion	$(CH_3)_3N$	9.8	Ethanol		16
Phenol		10.0	Cyclopentadiene		16
$O_2N{-}CH_3$ Nitromethane	$O_2N{-}\bar{C}H_2$	10.2	Propan-2-ol (Isopropyl alcohol)		16.5
Ethanethiol		10.6	Ethanamide (Acetamide)		17
$H_3C{-}\overset{+}{N}H_3$ Methylammonium ion	$H_3C{-}NH_2$	10.63	Methylpropan-2-ol (tert-Butyl alcohol)		19
Ethyl 3-oxobutanoate (Acetoacetic ester)		11			

(continued)

Values of pK_a for Various Acids[a] *(continued)*

ACID	CONJUGATE BASE	pK_a	ACID	CONJUGATE BASE	pK_a
Propanone (Acetone)		20	NH$_3$ Ammonia	H$_2$N$^-$	36
Ethyl ethanoate (Ethyl acetate)		25	N-Methylmethanamine (Dimethylamine)		38
N≡C—CH$_3$ Ethanenitrile (Acetonitrile)	N≡C—C̄H$_2$	25	Toluene (Methylbenzene)		40
HC≡CH Ethyne (Acetylene)	HC≡C$^-$	25	Benzene		43
Aniline (Phenylamine)		27	H$_2$C=CH$_2$ Ethene (Ethylene)	H$_2$C=C̄H	44
H$_2$ Hydrogen gas	H$^-$	35	Ethoxyethane (Diethyl ether)		45
Dimethyl sulfoxide		35	CH$_4$ Methane	H$_3$C$^-$	48
			CH$_3$CH$_3$ Ethane	CH$_3$C̄H$_2$	50

[a]pK_a = −log K_a. The less positive (or more negative) the pK_a value, the stronger the acid relative to another acid.
[b]In older textbooks, the pK_a values of H$_3$O$^+$ and H$_2$O are reported to be −1.7 and 15.7, respectively, but by definition they are 0 and 14.
Karty, J. *Organic Chemistry: Principles and Mechanisms*, 3rd ed.; W. W. Norton: New York, 2022.

Index

L

laboratory notebook, 25–32
  about, 26
  general guidelines, 26
  informal laboratory reports and, 34
  pre-lab flowchart, 28, 29–32, 30–32
  sections to be filled out pre-lab, 27–28
  section to be filled out during lab, 28, 29, 34, 37
laboratory reports
  about, 34
  evaluation, 37
  format, 34
  sections, 34, 35–37
λ (wavelength), 78, 79
λ_{max} (wavelength maximum absorption), 79, 80–81, 80–81
Liebig, Justus von, 136–137
life expectancy, 4–6, 5
limiting reagent, 27
lipophilicity, 10, 43
liquid-liquid extraction techniques, 43–48
  background, 43–44, 44
  examples of, 47–48, 47–48
  for obtaining organic products, 44, 45
  practical aspects of, 46–47, 47
  for separation of organic compounds, 44–45, 46
liquid sample preparation, for IR spectroscopy, 85–86
liquid–vapor phase diagram, 50–51, 51
longer-range coupling, 111–112, 112

M

magnetically active (spin-active) nucleus, 95, 95
magnetic anisotropy, 100, 101
magnetic dipole (μ), 79, 95, 95, 103
magnetic dipole transitions, 94
magnetic resonance imaging (MRI). See nuclear magnetic resonance (NMR) spectroscopy
magnetic sector analyzer, 128, 128
MALDI (matrix-assisted laser desorption ionization), 129
mass spectrometry, 125–142
  fragmentation and structure observations, 134, 134
  high-resolution mass spectrometry, 134–135, 135
  introduction, 127–128, 127–128
  ionization techniques, 129
  mass spectrum, 127–128, 127, 130–132, 131–132
  molecular formula determination, 132–134
  molecular formula overview, 126, 126
  practical aspects of, 130

sample preparation, 130
sample problems, 133–134, 135, 140–142, 140–142
mass spectrum, 127–128, 127, 130–132, 131–132
mass-to-charge ratio (m/z), 127, 128, 128, 132–133, 132
matrix-assisted laser desorption ionization (MALDI), 129
medical conditions, safety practices regarding, 17
melting point determination, 70–72, 71
mercury disposal guidelines, 21
(R)-(+)-3-methyladipic acid, synthesis of, 30–32, 30–32
mixed melting point, 72
mobile phase
  of chromatography (general), 60–61, 60
  in column chromatography, 67–68
  in gas chromatography, 61–62
  in thin-layer chromatography, 64
molar absorptivity or molar extinction coefficient (ε), 81–82
molar equivalent, 27
molecular bond vibrations, 78
molecular formula
  combustion analysis of, 136–138, 137
  confirmation of, 138
  introduction, 126, 126
  mass spectrometry and, 126, 126, 132–134
  sample problems, 137–138, 137, 139–142, 139–142
  structural clues from, 138–140
mother liquor, 55, 57
MRI (magnetic resonance imaging). See nuclear magnetic resonance (NMR) spectroscopy
multiplicity (signal splitting)
  in ^{13}C NMR spectroscopy, 120–121, 122
  data interpretation, 97
  data reporting, 116–117, 116–117
  introduction, 103–110, 104–108, 110
m/z (mass-to-charge ratio), 127, 128, 128, 132–133, 132

N

N equivalent hydrogens, 106–107
NMR spectroscopy. See nuclear magnetic resonance (NMR) spectroscopy
nonequivalent hydrogen coupling
  introduction, 97–98, 98
  multiplicity and, 103–104, 104, 106–107, 107–108, 109
nonhazardous waste disposal, 21
nuclear magnetic resonance (NMR) spectroscopy, 93–124
  ^{13}C NMR spectroscopy, 120–124. See also ^{13}C nuclear magnetic resonance (NMR) spectroscopy

data for laboratory report, 36
^{1}H NMR spectroscopy, 94–120. See also ^{1}H nuclear magnetic resonance (NMR) spectroscopy
  infrared spectroscopy comparison, 94
  introduction, 78, 94
nucleation sites, 56–57
nutritional supplements, 67

O

Occupational Safety and Health Administration (OSHA), 17, 18
oiling, 59
optical activity, 73, 73
optical rotation, 73, 73, 74
organic chemistry
  chemical property resources, 29
  costs of, 6–7
  green chemistry principles, 6, 7–12
  impact on humanity, 4–5, 5
  organic reactions, 4–5, 4–5
  overview, 3
  safety guidelines, 15–24, 29. See also safety practices
organic waste disposal guidelines, 20, 21
ortho relationship, 122
OSHA (Occupational Safety and Health Administration), 17, 18
overlapping signals, in NMR spectroscopy, 110, 110

P

partition coefficient (distribution coefficient, K), 43–44
Pascal's triangle, 105, 105, 111
penicillin, 4–6, 4
%ee (enantiomeric excess), 74
%T (transmittance), 85
personal protective equipment (PPE), 19–20
petroleum feedstocks, 11
phase (liquid–vapor phase) diagram, 50–51, 51
Φ (dihedral angle), 106, 106, 107
photodiode array detector, 79
π* orbital, 81
π bond(s), 80–81, 80
Planck's constant, 78
plane-polarized light, 72–73, 73
plant feedstocks, 11
polarimetry, 72–75, 73
polarity, gas chromatography and, 61
PPE (personal protective equipment), 19–20
precipitation, as recrystallization issue, 58
Pregl, Fritz, 137
pre-lab flowchart, 28, 29–32, 30–32
pre-lab preparation. See laboratory notebook